UNDOING OPTIMIZATION

Undoing Optimization

Civic Action in Smart Cities

———————————————

ALISON B. POWELL

Yale

UNIVERSITY PRESS

New Haven and London

Published with assistance from the foundation established
in memory of Calvin Chapin of the Class
of 1788, Yale College.

Yale University Press books may be purchased in quantity for educa-
tional, business, or promotional use. For information, please e-mail
sales.press@yale.edu (U.S. office) or sales@yaleup.co.uk (U.K. office).

Set in Janson type by IDS Infotech Ltd.
Printed in the United States of America.

Library of Congress Control Number: 2020943776
ISBN 978-0-300-22380-4 (hardcover : alk. paper)

A catalogue record for this book is available from the British Library.

This paper meets the requirements of ANSI/NISO Z39.48-1992
(Permanence of Paper).

10 9 8 7 6 5 4 3 2 1

For my mother and my daughter

Contents

CONTENTS

Preface

This book was written over a long period of time, between 2008 and 2019. In 2020, as it went to press, the COVID-19 pandemic unrolled across the world. In response to the differential impacts of this pandemic on Black people and people of color, combined with the systemic racism of police, justice, and health systems in many countries, researchers and activists began the deep rethinking and transformation of their work.

These transformations will likely change how you, the reader, encounter this work. The use of digital data to measure and control movement may feel commonplace, with limited capacity for challenge. Equally, the discussion of technical systems as if they were neutral and available for an abstract "citizen" to use may feel naïve in a context where techno-systemic power is understood to be used in reinforcing white supremacy.

I invite you to approach this book as a story of how we have arrived at this point, and to read its final chapters as my attempts to elucidate one of many possible ways to otherwise proceed. If the smart cities I describe here already seem quaint, this is because the dynamics that I sketch here are still in play—and there are ways to learn from them.

Acknowledgments

This book was long in coming. A project of over a decade amasses an enormous collection of people whose time, words, actions, and interactions have had influence. To anyone I have unintentionally overlooked here, as well as all those mentioned, thank you for making this happen.

Without Joe Calamia, who saw its potential and persistently shepherded it into this final form, this book would not exist. The advocates, activists, and practitioners who have shared their insights and projects with me have given their ideas to me to share. I thank them all, especially Juergen Neumann, Greg Bloom, Sara Heitlinger, Tim Davies, and Mara Balestrini for inviting me into their lives and projects (sometimes in conversations years apart) and Michael Lenczner for many years of travel together in the areas of community wireless and open data. Thank you to the local governments in Bristol and Fredericton for their openness. I also thank Leslie Shade, Andrew Clement, the late Michael Gurstein, Marita Moll, and Catherine Middleton for starting me on this path.

In my home department at the London School of Economics and Political Science I thank Robin Mansell, Lilie Choulairaki, and Sonia Livingstone for exceptional mentorship and Nick Couldry for kicking off my work on critical data studies. Moral, intellectual, and emotional support have come in particular from Myria Georgiou, Ellen Helsper, Seeta Peña Gangadharan, Bart Cammaerts, Omar Al-Ghazzi, and Lee Edwards, who also provided excellent comments on some early chapters. Jean-Christophe Plantin helped to develop some of the ideas in Chapter 2 and let me talk about them endlessly. Research assistance came from Arpan Ganguli, Karen Morton, and Svetlana Smirnova, and PhD supervisions with Svetlana and with Sebastián Lehuede, João Magalhaes, Henry Lyons, and Ville Aula inspired me. The students in the Data and Society program keep challenging me and improving my thinking year after year. Outside my department, Adam Greenfield, David Madden, and Judy Wacjman asked difficult and generative questions.

I'm grateful to my Vancouver colleagues for a productive sabbatical in 2017: the Information Studies Department at the University of British Columbia provided a quiet office and access to the Museum of Anthropology, Garnet Hertz provided studio space at Emily Carr University of Art + Design, and Peter Chow-White invited me into his convivial lab at Simon Fraser University.

Teresa Dillon, Mark Graham, Tracey Lauriault, Rob Kitchin, Christian Sandvig, Renée Sieber, Sharon Strover, Annemarie Naylor, and Linnet Taylor have been fantastic interlocutors. Intellectual life in London would be incredibly dull without the Usual Suspects—including Ian Brown, Jon Crowcroft, Jonathan Cave, Lilian Edwards, Hamed Hadiddi,

Chris Marsden, and Michael Veale. Working with Selena Nemorin, Funda Ustek-Spilda, and Irina Shklovski has illuminated ways to tolerate friction and practice care: I am so grateful to have learned with you. Gina Neff coaxed the early versions of chapters out of me and helped me structure the book. Alena Buyx introduced me to bioethics while we both held babies: her ideas and practice of care have stayed with me. My Saskatchewan friends Naomi Beingessner and Martha Poon helped me find ways to share my hybrid knowledge. Thanks to Will Hall for early encouragement and practical support, to Jen Marzullo and Joy Elton for holding me tight in London, and to the Aerial Family for keeping me fit and sane.

I thank my whole extended family for demonstrating so many ways to be creative, intellectual, and loving and my dad, David, and brothers, Eric and Nate, for nourishing a shared love of maps, cities, and creative experimentation.

Completing a book is all-encompassing, yet day-to-day life goes on. For the endless everyday acts of solidarity—large and small—without which my life as a scholar and parent would be impossible, I thank the women of Kennington, South East London. I couldn't have done it without you. Finally, and with all my love and hope for the future, I dedicate this book to my daughter, Hannah, whose joy and curiosity abound.

UNDOING OPTIMIZATION

Technology, Citizenship, and Frameworks of the Smart City

I'M standing with my six-year-old daughter, waiting to cross the road at the main intersection in my neighborhood. It's very busy; heavy trucks loaded with construction rubble are trundling out of Central London, double-buses are stacking up to collect passengers at the nearby bus stops, and private cars are zooming past, trying to avoid the inner-city ultra-low-emission zone, which costs £25.50 (around $33.50) to enter. We wait for long minutes until the green walk-and-bike sign turns on. Then we hurry across, because this signal doesn't usually last long enough for a child to walk across the road before it switches back to letting the cars, trucks, and buses through. When we get to the other side, my daughter asks, "How do they decide when the lights should change? Is there a person somewhere who does it, or is it a computer? Do they make them change at different times when there are not so many cars?"

The answer takes a little time to unpack. Traffic planners in London certainly employ real-time feedback from traffic flows and predictive algorithms to manage the phasing of traffic lights on the kinds of busy roads that we cross. This is an example of "smart city" design, where new (often digital) technology layers over and augments the spaces that are already there. The technology is one thing, but the design also includes ideas about technology and data that influence how it is possible to understand a city and act in it. The "congestion zone," for example, is a policy strategy that is intended to do two things: (1) discourage vehicle traffic in Central London and (2) raise money for reinvestment in the transport system. To do this, it links with the automatic license plate recognition system and the traffic violation system. The capacity for "smart" billing and pursuit of people who don't register their vehicles in advance of driving into London combines with the policy ideal of a less congested downtown and a supplementary revenue stream for a transit system whose government funding continually decreases.

SMART CITIES AND CIVIC ACTION

The intersection in my neighborhood bothers a lot of people, including many people with small children. Often a casual meeting will turn up a story of nearly being hit by a car racing through the light. Lots of parents, stranded on the sidewalk while the pedestrian crossing is blocked with backed-up traffic, look at each other in exasperation. Over the past several years various neighbors have started campaigns to advocate for the pedestrian light to be on longer. A petition was signed by several thousand residents using an e-petition that was circulated through several local Facebook and WhatsApp groups and then addressed to the local government. A neighbor who is well known in the local media contacted our constituency's member of Parliament. Others asked the local city councilors and were informed that vehicle traffic was expected to be "abnormally high" until the redevelopment of a nearby neighborhood was complete—in four years' time. Another neighbor contacted the police, who dispatched several community-service officers to observe the intersection for a few hours one day. Yet another set of neighbors undertook a separate action, supported by the local church, which had some of the local children demonstrate against the poor air quality caused by the traffic. A photograph of the children appeared in the local newspaper.

Some of this advocacy used networked online communication sources to bring together more people than might have signed a paper petition. Most of it, though, was fairly traditional— seeking media attention and using personal connections or political representation to raise awareness of the general issue of air quality (rather than the specific one of a congested or unsafe road crossing). What stops the people bothered by the light timings from being able to act effectively is a form of information asymmetry that is based on smart technology. The information on traffic flow is gathered by sensors. Quite likely, some part of the sensor system is owned by a company that collects the data, analyzes it, and passes on the analytic insights to Transport for London. Pedestrians don't know whether information is gathered about their own movements or only vehicle movements (the latter would make sense because of the license plate cameras). The neighborhood activists don't have a count of people, accidents, or near-misses that is comparable with the data on traffic flow. Perhaps the next step for the local activists should be to install their own sensors to measure how many people are stopped from crossing the road or to measure pedestrian congestion at different times of the day. If they did, they would join the kinds of people I write about in this book: people who notice how companies and governments talk about networks, data, and sensors and who try to respond using the same kinds of technologies—and arguments.

In this book I explain how it came to be that priorities for traffic are driven by digital data from sensors and why many citizen groups feel that their stories are best communicated in the same way. To do this, I put together a history of the idea of the smart city and explain how technologies and ideas about citizenship have changed, with new visions of "smartness" layering over each other. What began as an idea about improving citizens' access to knowledge by expanding access to the internet built up into a set of systems oriented toward extracting, modeling, and optimizing systems based on data. In turn, a focus on the value and legitimacy of data suggested that "sensing" the city might create another form of smartness. In each of these modes, citizens have had to come to terms with different idealized roles for their actions. As for my neighbors, these idealized roles suggest the possibility of participation in the function of the smart city but leave it unclear what power this kind of participation entails. By trying to understand what the smart city was, is, and is becoming, it is possible to see how citizen power, government power, and corporate power are shifting.

WHAT'S THIS SMART CITY?

In the past decade, the dominant vision for the smart city has shifted from focusing on a communicative city equipped with access to information toward an idea of a pervasively connected city where data can be activated to increase efficiency. This move, from technologies of access to technologies of data, works to support a vision I call the "big data optimized city" because it depends on seeing connectivity not as an end in itself but as a precondition for the production of data that, in turn, can be processed to generate efficiencies—to "optimize" transport, city service delivery, or the exercise of civic participation. Like access, optimization is an ideal that transforms the arrangement of technological and social resources in a city, laying down technology-driven assumptions about how social life should unfold.

In the following chapters I show how these technology-driven assumptions came to take center stage in framing certain forms of civic action. In the fields of communication studies and science and technology studies, these ways of thinking about, building, and structuring systems are referred to as technosocial imaginaries. This term has its origins in the work of philosopher Charles Taylor, who describes how a society's practices begin to make sense to the people who participate in them. This process of making sense is what he calls a "social imaginary."[1] Taylor argues that almost all of what we consider to be modernity is built up from a way of thinking that first circulated among influential groups of people and then became widely accepted. This way of thinking included the idea of the rational individual who was able to claim rights.

I consider that one part of society's making sense now has to do with the place of technology in sustaining this broader social imaginary of modernity. Like Taylor, I also observe how certain understandings and expectations about the way that the world should be are built up by influential entities and thereafter begin to be taken for granted. However, I also work with an idea that another scholar influenced by Taylor has developed. Robin Mansell's idea of the imaginary integrates the idea of modernist world-making that Taylor began with, but focuses on the spaces where technologies can be imagined or reimagined. Mansell argues that technologies like the internet can sustain a "dominant" or, conversely, an "alternative imaginary" and that these build up expected realities in part by creating dominant ways of claiming power and establishing value—often economic value. In the case of the internet, a dominant imaginary focuses on the generation of value through the creation of intellectual property, and an alternative one claims social and public value through the maintenance of the internet's openness and accessibility. Groups of actors (individuals, organizations, and institutions) build up their claims on either side by referring to specific and assumed-to-be-consistent features of the internet: its ability to circulate easily reproduced digital information and its networked architecture. The important point here is that sociotechnical imaginaries are not mere visions; they are sustained also by the creation and maintenance of technological systems and by the alignment of particular ideas about how things ought to be with what technologies have made possible.[2]

This is where the smart city comes in. What I show in this book is that a particular imaginary about how to manage urban complexity has built up in and around expectations about technology. There is a long history of abstracting the function of the city and trying to understand it as if it were a system. Here, I focus on the particular technosocial imaginaries that have built up around the assumptions that cities as systems can be understood through, and improved by application of and development in relation to, technological systems. I also look at how ways of imagining cities as technological systems has influenced the possibilities for citizens to act within them. In particular, I hope to illustrate how what I call techno-systems thinking remains consistent across a number of different technological promises for the smart city, while also explaining how the big data optimized city has emerged as a particular version of the smart city in an age where data extraction and intermediation have met predictive decision-making and risk management.

Building on the ideas about the construction of social imaginaries, I show that the key principles of techno-systems thinking are built up by influential actors like technology companies but live on in all kinds of projects and efforts, including ones intended to democratize, open up, and remake the smart city for citizens. To show how this happens, I report from smart-cities projects around the world, presenting the views of people working on them. I examine community-based wireless networks, open data strategies, and citizen-sensing projects. Here, in the words of participants, I describe arguments similar to those made by technology companies. I observe attempts to employ—either generatively or disruptively—the available technological systems, as well as the arguments and discourses that give these systems social legitimacy.

My travels between these different projects, undertaken over a period of ten years, has showed me how the imaginaries introduced by a small number of influential actors spread, becoming part of the reality of city life. Because these imaginaries are technosocial and specifically focused on the significance of systems, different groups of people can take them up by, for example, trying to influence city decision-making by collecting different kinds of sensor data, building a self-run wireless internet access network, and describing work on it as "community-building." All of these initially appear to be alternative imaginaries. However, the evidence set forth in this book suggests that creating a strong opposition between dominant and alternative sociotechnical imaginaries may not be serving us well. What observations and analysis of smart-city projects show is that the world-making potential of techno-systemic thinking winds its way even into the definitions and potential for civic action. Advocates, activists, creative technologists, and even buildings designed as citizen-oriented smart-city test beds cast their potential interventions in relation to language, practice, and system architecture drawn from the repertoire of technology companies. Equally, technology companies and urban policy-makers appropriate the systems and modes of thinking of those generating alternatives—whether the alternatives are distributed networks making access to technology possible or whether they are ideas of civic data collection.

BIG DATA OPTIMIZED CITIES AND HOW WE GOT THEM

Expanded capacities for data collection, along with interest in reducing risk and predicting disorder, have become influential aspects of a vision of the smart city that depends on employing data to map and understand urban populations and to generate value by selling, trading, or analyzing it. One of my key goals in this book is to describe the genesis of this particular techno-systemic imaginary of the smart city. Another is to describe how it has intensified and with what consequences. The first goal necessarily involves identifying the technosocial conditions under which this big data optimization emerged.

The evidence presented here indicates that one factor enabling data collection and analysis on a broad scale is the presence of existing infrastructure for networked connectivity. This infrastructure has itself been the subject of contestation, reimagining, and mutual appropriations of different modes of techno-systemic thinking. Yet what has resulted, as I detail in the early chapters of this book, is a contingent and contested space that has yet to resolve into a true alternative to the big data optimized city.

A second goal requires a focus on where intense datafication might lead. Curiously, one direction forward from the focus on collecting data, information, or "evidence" as a key practice in the city is to extend the kinds of data collected outside ordinary realms of human experience. Observing the world through the use of environmental sensors contributes to data-informed civic conversations and extends the knowledge of urban spaces into new realms. Sensing the environment, animals, and urban experience in new ways may prove to be a generative counterpoint to a focus on optimization.

In short, the data-based smart city comes from somewhere and has the potential to go somewhere unexpected. To lay out this genealogy and possibility, I focus on the relationships between networks of communication, data for optimization, and sensing the other. These relationships unfold across the coming chapters and are illustrated through presentation of results from empirical work on relevant smart-city projects. My engagement in these projects has varied, from working on a long-term ethnography of wireless access projects to conducting focus groups and interviews with open data advocates and having more systematic interviews and analysis of project documentation of environmental sensing projects. The value of including observations and insights from people working on these projects is that they reveal the contingencies and complexities that are inevitable consequences of dealing with techno-systems thinking.

All of these projects have also engaged with the structural realities of how people who live in cities are able to act, especially in relation to the often neutral or abstract technologies of connectivity or data. In the past twenty years, just as international migration and neoliberal governance have transformed the populations and power structures of cities, different forms of civic action have been made possible in relation to these technologies, drawing from and connecting with broad shifts in work, organizations, and capitalist power. In the overall context of expanding cultures and practices of neoliberalism, the relationship between state and citizen has shifted from "command and control" to coercive governance. We can understand governance as a set of power relationships that depart from command and control and feature more coercive modes of influence, including building expectations for partnership, multi-stakeholder collaboration, or participation that extract energy from participants, provide illusions of influence, but do not disrupt deeper modes of power.[3] To illustrate, think of my neighborhood activists, who have made their voices heard about their view on the road crossing, but are still not able to understand how decisions are made about it and whether these can be changed.

Such forms of governance are understood to involve people in disciplining themselves and in internalizing expectations about how they ought to behave. Sociotechnical imaginaries, like the idea of techno-systems as generative of urban order, play important roles in governance processes by making some kinds of realities and actions visible and present. We can observe how engagements with these forms of governance build up over time, as ideas of smart cities based on networks of communication set the infrastructural and social groundwork for ideas of smart cities as based around the use of data for optimization, and how these in turn opened out perspectives that acknowledged the expressive capacities of data and, in turn, built a space for smart cities concerned with sensing the other.

NETWORKS OF COMMUNICATION

In Chapter 1, I return to an early technological framing of the smart-city concept, looking at how companies, governments, and groups of techy citizens lay claim to the power of connecting cities—and people—to the internet. I show how the language about networking employed by Cisco, one of the leading providers of wireless internet access technology, shaped and was also influenced by claims made by activists about a "right to communicate"—including claims about communication rights as central to the governance of communication systems. The right to communicate—an abstract concept mobilized by scholars and activists much the way David Harvey makes a claim about the "right to the city"—connects with the idea that communication is a fundamental exercise of citizenship: becoming informed, participating, being heard. As a first step toward exercising a right to communicate, a citizen needs some means of accessing information and being heard.[4]

Here is where the right to communicate connects with the idea of the smart city and with a certain fantastical set of claims about networks. In the late 1990s and early 2000s social scientists were beginning to investigate how connections between people and institutions generated certain forms of power. Suggestively, some thought that a massive social change was afoot and that existing modes of social organizations, like hierarchies at work and the distribution of value in economies, were being transformed into more horizontal, interconnected spaces.[5] Sociologist Manuel Castells made an explicit connection between the power of the network as a social form and the networked architecture of the internet: in *The Internet Galaxy*, first published in 2002, he described how internet connectivity intensified a separation between what he referred to as the "space of flows," linking interconnected, economically influential locations, and the "space of places," which were less influential, less connected locations. Castells's work arrived at roughly the same time as the commercialization of the internet and the first dotcom boom. Neoliberal economic theorist Thomas Friedman evoked the same concept in economic terms with his book *The World Is Flat*, in which he argued that interconnection, accomplished in part with networked communication technology, would lessen the influence of distance and even out economic opportunity.[6] Excitement about the potential of increased access to information led to debate about how this access should be provided, which is where the idea of communication rights reappears.

Two of the first stories in this book are from Canada, where national policies that developed in the late 1990s used communication rights as a rationale for providing public support to access to internet infrastructure. The same rationales, however, were also used by Cisco and other technology companies. Language affirming the power of networks as self-evident appears in government officials' descriptions of the significance of a wireless access network funded by Cisco and installed in a small city in Canada's eastern province of New Brunswick.

Activists were also influential in these spaces. The idea of communication rights connected with enthusiasm about the new possibilities created by internet access and access to open-source technologies. Network architectures inspired certain types of political claims, as did activist actions like those of the "free wireless" movements that built wireless internet "mesh networks" using interconnected radios while using these technical projects to illustrate a set of political claims. This concatenation of technology and political argument also appears with a group of young people in Montreal who explicitly claimed that their volunteer-built wireless internet access network was "hacking the built city." These young people's engagement with open-source technology similar to that marketed by Cisco and with policy frameworks of communication rights became enfolded in broad, worldwide enthusiasm about wireless networking.

In these threads and negotiations we see consistent techno-systems thinking: the imagining of specific capacities for networks.[7] Here, networks link up disparate individual points to create a new whole. The space for activism is outlined by the possibility that citizens might be able to build their own networks and by the way that the rights to communicate might be connected with rights to practice, exercise, or explore the city in relation to what technology is understood to afford.

The notion of the internet as a space for communication has been co-produced along with the notion of a city as a space for the exercise of expanded socioeconomic and cultural rights, some of which could be exercised through certain kinds of participation—including consuming the information delivered over the internet, producing demand for this information, and even (for the activists making their own network) creating the very infrastructure delivering it.[8]

As internet access became pervasive across many cities in the Global North, the notion of smartness shifted. Accompanied by some of the same language and enthusiasm, a new version of smartness emerged that focused on the value and significance of data. As with the issue of access to communication infrastructure, access to data became framed as a space for citizen action. However, rather than being embedded in policy advocacy connected with rights claims, this space was demarcated by political claims about the importance of civic action in the context of smaller or less powerful government and tied to technology industry ideas about the value and importance of platforms as digital information intermediaries.

DATA FOR OPTIMIZATION

The appearance and significance of data as an organizing context for the smart city depend on a combination of shifts in sociological and technological frames. Datafication is now understood canonically as a process that transforms many aspects of everyday life into data, and data into value, and it is considered a "legitimate means to *access, understand* and *monitor* people's behavior."[9] Datafication also changes how communication is understood. In a departure from the idea of "speaking, listening, and being heard" that shaped enthusiasm about access to the internet, datafication suggests that data production and circulation are significant in themselves, regardless of what the data might contain. This parallels an observation made by cultural theorist Jodi Dean about the shifts in expectation relating to participation that emerge as part of "communicative capitalism." Here, information, communication, and participation become norms of publicity rather than the means to the political ends that they were presumed to serve.[10]

These features of datafication are important because they indicate changes in the expectations around how citizens are able to generate and structure their participation. Networked communication technologies "shape our perceptions of reality more comprehensively" than other artifacts because they are "tools for producing useful results and tools for representing the world." A contribution to Twitter can be a form of personal expression or a data point in an analysis predicting future demand for services—or the potential for illegal activity.[11] Depending on how networked communication technologies are architected and governed, they can constrain or enable emergent forms of citizenship. They can facilitate surveillance and strong top-down control or strengthen ad hoc relationships.

Data-based technologies, too, have been the cause and effect of techno-systems thinking about smart cities. Because raw data is not inherently valuable, some of the key struggles related to data citizenships involve how commercial (or sometime civic) intermediaries perform the analytics that are meant to make meaning out of data. The shift is toward not only datafication within smart cities but forms of governance concerned with data aggregation, risk reduction, and the generation of prospective suggestions for future forms of control. I identify how these aims are associated with the shift toward "platform governance" in cities, where power, decision-making, and information resources are brokered by intermediaries. Similar to the process of urban "splintering," whereby physical resources in cities are privatized, informational platformization also reorders important resources away from the public realm. The big difference in the construction of this big data optimized version of the smart city is that the power shifts occur in a way that builds upon, rather than exemplifies, the neoliberal governance that I described earlier.

Something in this platformed big data optimized imaginary of the smart city goes beyond the neoliberal splintering, as I explore in Chapter 2. While splintering separates out resources and partitions the city, optimization instead tends toward celebrating and valorizing control of entire systems and limiting citizen action or civic participation to the kinds of contributions that make the systems work.

Put another way, datafication opens out spaces for citizens to participate in by generating data, auditing government data, or creating alternative data analytic systems, but these spaces for action still follow the dominant logic of big data optimization. I look into these dynamics in Chapter 3, which includes reports gathered through interviews and focus groups with open data activists in the United Kingdom. I treat the open data movement as an important case in examining how political and commercial understandings of datafication, platform governance, and optimization shaped the space occupied by citizens wanting to use open data.

This space also turned around the responsibility for citizens to enact values like transparency and openness, values often associated with neoliberal forms of citizenship. This turnaround is illustrated by an excerpt from the website of the Open Data Institute (ODI), a UK think tank: "Our open, peer-to-peer culture is fostering a new generation, 'Generation Open,' who can use open data, tools and systems to transform our architecture, our environment and our society for the better."[12] Such an open society is also sometimes given as a goal, which motivates activists to critique open data. The actions are ambivalent: open data advocacy can disconnect data production from the consumption of civic services, and critique of data systems is sometimes used to celebrate the shrinking of government and the importance of platforms.

With more platforms and more discussion of the importance of data, a third techno-systemic framework comes into view. This one starts to orient toward sensing and the possibility to "speak" and "be heard" using digital devices that collect environmental information.[13] Raising questions about who sensing is for, how data should be managed, and who might benefit, the introduction of sensing into the space of the smart city also suggests more philosophical questions about urban life. After all, sensing is one way to encounter other forms of knowledge.

SENSING THE OTHER

The final techno-systemic way of imagining the city that I examine here is the idea of the smart city as a sensory space, where data-collecting technologies model the seen and unseen in real time, bringing together measures of air quality, noise, humidity, temperature, and the movement of animals. This vision of the smart city has a substantial history within fields of urban design and computation, which often celebrate the dynamic and interrelated forms of knowledge that can emerge from expanded sensing regimes.[14] It is true that sensing provides ways of knowing that are different and dynamic and take into account experiences whose scan and scope might otherwise make them ungraspable. However, sensing also intensifies the focus on and efforts to control risk that the big data optimized smart city builds up. Sensing cities can therefore be surveillance cities, where ideal citizen-subjects consent to data collection in exchange for the privilege and pleasure of improved information and the certainty of well-managed urban systems. Meanwhile, alternative imaginaries of sensing cities demonstrate the possibilities for new kinds of ethical relationships predicated on the ability to represent city life as an ecosystem. This might mean accepting that optimized cities fail to account for how knowledge in places changes dynamically over time—how it hybridizes.

In Chapter 4 I investigate the gap between what sensor data can be made to say, that is, who it might speak on behalf of, and what institutional and social contexts this voice might connect with. Tensions result from (and are inherent in) the creation of a civic-sensing pilot project to address issues of unhealthy damp in buildings, part of a smart-cities pilot created in Bristol, a city in the west of England. Developed as part of a bottom-up community-led effort to identify what kinds of problems civic sensors could solve, the pilot created sensors to identify damp buildings as part of a civic strategy to put pressure on the government to provide good-quality housing. Ironically, the sensing in part replaced in-person damp inspection after the latter was deemed too costly for the local government. Participants in the project, including government collaborators, point out that sensing and measuring, even when they are intended to fill gaps and enhance civic voice, cannot themselves account for missing people or institutional processes. Here is where I begin to suggest that perhaps the civic contribution of sensing might be better captured as inevitably partial, in tension, and unstable, and that the frictions could inspire new forms of solidarity in connection with collected sensed data.

The evocation of an "unknowable city" partially and contingently revealed through sensing sets up the approach taken in Chapter 5, where I discuss other ways to consider the influence of sensed data. We can take sensing as an invitation to reflect on what there is to know about urban life and what ways this knowledge presents in the contexts of contemporary global cities, where many different kinds of "others"—people on the move, as well as plants and animals—experience the city in ways that might never be fully known. We can usefully explore philosophical perspectives that might provide some capacity to rethink the relationship between knowledge, location, and experience by focusing on a smart-city pilot installation of sensors in a community garden tended by master gardeners who arrived in London from many other parts of the world. Examining the frictions produced in this project opens out the space of the smart city to include not only the gaps and frictions inherent in sensing citizenship but also the radically contingent relationships between people, plants, and collective spaces—in this case, the space of a community garden.

As anthropologist Anna Tsing writes, "Human exceptionalism blinds us. Science has inherited stories about human mastery from the great monotheistic religions. These stories fuel assumptions about human autonomy, and they direct questions to the human control of nature, on the one hand, or human impact on nature, on the other, rather than to species interdependence." Paradigms of human management and control fail to account for the reality of interdependence, and indeed most smart-cities sensing projects treat animal interaction as an error. The designers of a water-quality sensing system in Oxford, for example, note that one of the limitations of their system's capacity to raise alerts of foreign bodies in the water is that "it cannot tell the difference between a goose and a shopping cart."[15] Similarly, designers of a smart-city sensing test bed in Glasgow report that one of the initial issues with its motion-sensitive security monitors was the way that the nocturnal movements of foxes triggered the sensors designed to monitor suspicious human activity.

These "errors" highlight how other species appear in current sensing regimes. For urban dwellers, who have thoroughly internalized the idea that "our species being is realigned to stop Others at home's door," the most radical potential of sensing networks may be in understanding and needing to accept the connections between what is like us and what is foreign: a kind of broader empathy that is essential for understanding relationships between people who are others to each other.[16]

CITIZENSHIP AND THE COMMONS

Across all of the chapters of this book, the idea of citizenship resonates, as does the idea of the commons. Both of these ideas are built up in relation to the particular version of the smart city. The commons, in particular, acts as a kind of persistent civic or bottom-up alternative vision for how communicative resources can be generated, stored, or used for collective value.

The concept of citizenship I employ is both broad in reach and narrow in application. Residents of cities may not be citizens of the nations the cities are located within, and in sanctuary cities they may not even be legal residents of the associated nations, but they may participate economically or contribute to local services through taxation, contribution to voluntary organizations, or participation in local politics.[17] This broad concept of citizenship contrasts with a narrow focus on how civic action unfolds when it is specifically conceived of in relation to technology. The narrower notion creates unstable expectations of civic responsibility. In the early chapters of this book I show how responsibility and citizenship have been positioned in terms of rights (including the right to communicate), and in the later chapters I show how this view has shifted to identify citizens as consumers of city services optimized by big data analytics or to celebrate them as "auditors" who hold public officials to account. The result can be, as I explicate, that certain people, practices, and modes of civic action are more visible, valorized, or facilitated by the frames of smart citizenship. These ideal enactments of "good" or "ideal" citizenship then begin to align with the ways that urban smartness is understood. If urban smartness is in line with the dominant techno-systems thinking, it is then complicated to claim that civic action in smart cities, especially action that takes place using technology, is a straightforward form of resistance.

I use the figure of the "commons" as a way of identifying how these different techno-systemic imaginaries cohere around the promise of spaces where benefits can be shared among the people who contribute to the joint resource. Elinor Ostrom, the economic theorist of the commons, included the internet as one example of what she defined as a "'common-pool resource'—a natural or man-made resource from which it is difficult to exclude or limit users once the resource is provided by nature or produced by humans."[18] The commons that appear in smart cities not only are human-produced networks from which it is difficult to exclude people but can also be information commons. One of the central considerations of how information commons should be managed is "the institutionalized community governance of the sharing and, in some cases, creation, of information, science, knowledge, data, and other types of intellectual and cultural resources."[19] In the early stages of smart-city development, these questions of governance applied to the management of access to the internet, foregrounding questions of equity and rights as the basis of citizenship, and in the later stages they have referred to a commons of data produced by citizens. This shift is important in our understanding of how citizens are invited to find space in the smart city and also important in exploring what can be done with the idea of the data commons. In the final section of the book, I suggest some radical ways of considering the city as a commons co-produced by many others. I use the idea of sensing urban plants and animals as a way to figure the many forms of difference that contribute, with significant friction, to the ways of knowing the city. As the notion of humans as the only important inhabitants of the world falls apart, these differences, frictions, and continued being-in-commons are essential to consider.

OPTIMIZATION AND OTHERS

How and to what extent can the structures put in place by the dominant actors within a capitalist techno-system be effectively reappropriated? Is reappropriation possible at all? There are clearly spaces of appropriation, but I question the extent to which they are resistant and ask whether we might need to examine civic action in relation to technology from a new point of view.

In the shift from communicative cities to ethically and environmentally sensitive cities, citizenship has both narrowed and broadened in relation to communication technologies. As normative concerns about communication rights have been juxtaposed with calls for transparency, and as that transparency has become buried in the process of managing risk, ideal neoliberal citizenships are proposed that support economies of data analysis and processes of system optimization and risk reduction. At the same time, other kinds of citizenships and other modes of governance also emerge. While alternative communicative, data, and sensing citizenships are not always capable of surpassing the dominant forms, they promise other ways of being in cities and perhaps other ways of being in the world.

Optimization may appear to be the natural motivator for the development of technological systems, but as my story about neighborhood traffic illustrates, not everything can be optimized. This book shows not only how the big data optimized smart city emerged but the way that its orientation toward risk reduction, prediction, and integration emerged from the success of previous activism directed at expanding claims to communication rights. Understanding the smart city means understanding how data-collection processes build on top of networking, creating expectations of good citizenship, and how these expectations open the space both for rethinking what kind of knowledge sensing might open up and for identifying how what we know changes all the time, hybridizing as it moves into new contexts and absorbs the capacities put forward by new technology. Finally, this book questions how these more expansive forms of knowledge influence ways of thinking about citizenship against the backdrop of techno-systemic optimization.

Citizens take up new technologies and use them to transform the possibilities for communication and the forms of life in cities. These appropriations, however, are not always resistant, and they often contribute in unexpected ways to perpetuating and intensifying certain dynamics of the smart city. Untangling how these transformations have occurred across different technological and policy contexts demonstrates both the new constraints placed on citizens and the urgent shifts in our ways of understanding urban smartness that are required for the current age of intensive urbanization, inequality, and climate emergency.

To go back to the intersection by my house: not only does it present problems to be solved (including those that can be easily optimized), but it also has incredible complexity. It looks different from my place on the sidewalk, from the cab of the heavy hauler routed down the A2 to a construction site, to my daughter and the other kids trying to run across the street, and to the pigeons darting overhead. As soon as the perspective on the problem narrows, as soon as the problem becomes singular rather than multiple, the power that collects around the entities defining techno-systemic imaginaries grows. Not only do corporations and governments gain power in the smart city. As I show, the ways they develop and build technological visions and practices also shape the space that we all have to traverse.

CHAPTER ONE

Network Access and the Smart City of Connectivity

HE smart city isn't new. For at least the past two decades, new communication technologies have been imagined, marketed, and constructed to improve the function and experience of urban life. Let's revisit the smart city of the late 1990s and early 2000s, when access to internet technologies was an important part of how smartness was imagined.

In many places in the world, governments and companies proposed projects to expand access to the internet. Often the entities pursuing these projects argued that expanded access to communication resources would lead to an expanded capacity for participation in civic life. Enthusiasm for projects like the Blacksburg Electronic Village—which aimed in the late 1990s to give the entire population of the town of Blacksburg, Virginia, access to the internet—turned on the idea that networked information systems could enhance participation in local decision-making and strengthen the ties between people through the benefit of enhanced connectivity to an information network.[1]

Of course, the actual experience for the beneficiaries of this expanded communications access departed from the vision. The early 2000s visions of internet-based democratization of civic action paralleled other visions of augmented cities as ideal sites of expanded connectivity. Both ideas connected with struggles over communication rights but also opened space for information technology (IT) or information and communication technology (ICT) companies to market various internet access technologies as pillars of urban smartness. Governments, companies, and activists defined—as well as created—different capacities within the idea of the smart city. The proposals wove together a promise of a particular kind of "technological sublime" that, in the early 1990s, referred to a networked set of relationships modeled on the real and imagined capacities of networks of internet access. Social theorist Leo Marx and communication scholar Vincent Mosco both identify the idea of a technological (or digital) sublime as a major mythology connected with the power and influence of technology.[2] Mosco argues that in order to judge the power of information technology we must also take account of the influence of major mythologies associated with it. The visions of smart cities derive from a myth about technology in general and optimization of civic life in particular. In practice, these kinds of visions developed through the language used in marketing by technology companies and in the investments that followed. They have echoes in public policy and in projects undertaken by technology activists and advocates, who often use the same discourses and the same technical materials to develop critical alternatives to the industrial visions for smart cities. This particular techno-systems thinking sets up expectations for the role of computer networks and internet access in cities.

INFORMATION SUPERHIGHWAYS, CONNECTED CITIZENS, AND 1990S SMART CITIES

Beginning in the mid-1990s, scholars began to understand the global city as a communication nexus. Global connections through the internet intensified the capital and influence of global cities, and flows of people produced different global citizen experiences, including experiences of cities. A way of describing how technological modes of thinking influenced experiences of urban space was the notion that connectivity created a separation of global "spaces of flows" and local "spaces of places" bounded by local experience. This distinction seems, in retrospect, too rigid to account for the ways that people, rich and poor, moved into and between cities in this period and produced new forms of citizenship related to changes in economic status.[3] The new economic relationships encouraged rights claims but also created new sites of civic struggle as migrant labor and transnational capital clashed. They also made the terrain of communication one of the sites of social struggle.

As claims for communication rights expanded, the result was a sense that within global cities, citizenship involved "more than rights to participate in politics. It could also include other kinds of rights in the public sphere, namely, civil, socio-economic, and cultural." In this more expansive but contested space for claiming rights, global and transnational linkages inspired by the promise of better internet connectivity were viewed as ways to develop new citizenships.[4] This view took the extreme form of claims that the institutional and cultural rigidities of the city might be liquefied or transcended by the more fluid, global, internetworked virtual world and worries that the space of the traditional city was declining, making "authentic urbanity" impossible and limiting the ability of citizens to enact democratic rights.[5]

The ideas of global network development, of the configuration of place-based relationships in terms of nodes and links, predated the expansion of information technology networks but became influential when interpreted in relation to the technical networks. Interpreted in a particular way, the identification of the space of flows serves as a justification for associating communication rights with access to internet services, and the celebration of this space of flows in counterpoint to a disconnected space of places justifies placing attention on connectivity and technological equipment as for citizenship rather than less networked modes of civic action. This is the first version of the smart city that I interrogate: the communicative city, where attention to networked modes of power across society and interest in increasing access to networks intersect with calls for communication rights. Citizens may have communication rights anticipated or planned for by policy-makers, or they may have them interpreted by companies as new opportunities to consume. When a smart city is focused on network connectivity, people can also participate in claiming communication rights by self-organizing to create different means of access to networks.

ARE COMMUNICATION RIGHTS SIMPLY
RIGHTS TO NETWORK ACCESS?

When the internet was commercialized in the 1990s, discussions of communication rights had already been under way for fifty years. The UN Declaration of Human Rights includes communication rights within its definition of cultural rights but omits the rights to freedom of speech as well as rights to seek information, generate ideas, express and speak, listen and be heard, understand and respond, learn and create, which are considered by some scholars as essential to a broader framing of communication rights.[6] This is where a significant tension emerges, one that will be important in this book. The space of information seeking, sharing, and creation has often been described as a commons—a space of collaboratively generated value where individual contributions are gathered together to create a shared resource of value to all who contribute to it. Again and again, the notion of a commons is evoked directly or indirectly as a normative ideal worth aspiring to. A communication commons promises meaningful engagement, participation, and deliberation—practices associated with the exercise of other types of democratic rights. In the late 1990s, it still seemed that increased access to information infrastructure might spur the development of this commons, although scholars like sociologist Michael Gurstein warned that this kind of development was based on the assumption that people were equally capable of making "effective use" of these technologies. At the same time, commodification of the space of communication was already under way, creating and intensifying what Jodi Dean describes as communicative capitalism, where "values held as central to democracy take material form in networked communications technology." The value of political participation is replaced with the promise of political speech via access to communication technology. Communication theorists Sonia Livingstone and Peter Lunt call the exercise of civic participation under these conditions "consumer citizenship."[7]

Arguments made in the early 2000s about the benefit of expanding access to the internet after its commercialization evoked the discourse of rights and the promise of increased democratic participation. However, as scholars noticed at the time, policy decisions also began to highlight access to goods and services rather than seeing ICT as a way to "foster and nurture participation," as communication scholar Leslie Shade put it. For example, in 1996 the Canadian government advocated for a "national strategy for access to essential services." By the early 2000s, with a Conservative government in power, the policy reframed citizens as consumers of communication services. One of the ways that these strategies and their reframings operated was through policy support for smart-city projects.[8]

The promise of "augmentation" of cities with network access technology, proposed by technology companies, captured the notion of communication rights and parlayed them into a justification for investment in internet access technologies. Long before the criticisms of the top-down formulations of smart cities and the disavowal of citizen needs, technology companies linked their promises to rights-based policy frameworks—setting them up as ideas that advocates and activists needed to contend with. In this way, dominant perspectives on the significance of access to communications for social benefit begin to define social success in technological terms. Such an "exogenous" ideal of development is divorced from local context and therefore more likely to be tied to success as defined by companies making profits.[9] What I refer to as techno-systems thinking, which is formed by this conceptual apparatus, can be glimpsed across commercial, policy, and civic language.

Smart-city models have always tended toward the exogenous and the abstract. They imply, as urbanist Adam Greenfield and Nurri Kim have put bluntly, that "there is one and only one correct solution to each identified need; that this solution can be arrived at algorithmically, via the operations of a technical system furnished with the proper inputs; and that this solution is something that can be encoded in public policy, without distortion."[10]

DEFINING THE SMART CITY: CISCO AND THE VISION OF WIRELESS EFFICIENCY

Over the past twenty years, companies like Cisco have defined smart-city perspectives, generating influential visions and discourses in the process. Cisco builds and markets wireless radio equipment, including the routers that make Wi-Fi internet connectivity possible. In the early 2000s, Cisco developed technology that was quickly appropriated by not only individuals but also groups who wanted to extend internet connectivity using radio waves. Because some routers used open-source software, they were especially interesting to hackers and tinkerers who reprogrammed them to transmit in new ways and often joined with community organizations to share internet connectivity within communities. Before 2005, most public Wi-Fi projects were local projects undertaken by small-scale organizations: neighborhoods, community organizations, and municipal governments. Beginning in the early 2000s, hundreds of cities and towns across North America began to invest in Wi-Fi projects, many based on public-private partnerships.[11] In the early 2000s, Cisco and other technology companies became involved in identifying test beds and models for technologically "connected cities"— Cisco's particular branding of the smart-city vision. To capitalize on the expansion of Wi-Fi and Cisco's manufacture of Wi-Fi antennas, the company developed a vision of a connected city and rewarded city governments who used their technology in demonstration projects, including a series of Connected City prizes. Below I describe a visit to one of the prizewinning cities, Fredericton, Canada, in 2007. The city government's adoption and interpretation of the benefit of this prize illustrate how rights claims about internet access, corporate promotion of connectivity, and public service provision intersected in a small city in a Canadian forest. As I discuss, the same ideas were developed in a more radical vein by activists who used the architectures of connectivity to model their version of the communication commons. Such projects took place in Berlin and Montreal at about the same time that Cisco also invested in creating the language and architecture of connected smartness.

CONNECTED CITIES AND THE VISION OF MUNICIPAL WIRELESS CONNECTIVITY

The small city of Fredericton, New Brunswick, located in the middle of the eastern Canadian forests on the Saint John River, pitched for Cisco Connected Communities funds in 2005.

They received Cisco support for a municipal Wi-Fi network that was touted on its completion as the first of its kind anywhere, providing broad wireless internet coverage of a downtown business district, including all offices, hotels, and businesses, free of charge and with connectivity bandwidth provided by the local government's fiber backhaul network. The case of this small city and its position as a celebrated example of a smart, connected city in the early 2000s illustrates the expansion of communication rights and discourses of "intelligent" or "knowledge" industries combined to position communication technologies as central to a "connected community." Although the local government made investments in broadband connectivity using connectivity and basic infrastructure as key metaphors, a certain ambivalence also characterized the way that connectivity was considered: primarily as a driver for economic expansion and not necessarily as a service to be extended to the most marginalized.

In 2007 I spent several winter months in Fredericton, based within the municipal government offices alongside the small team that designed and built and maintained the Wi-Fi network.[12] In my weeks in the snow-bound city I interviewed the main decision-makers within the local government, as well as tech-based business owners, university and industry researchers, and many other people whom I met in restaurants, on the ski trails, and at the weekend farmers' market. Why build a free Wi-Fi network in a small town surrounded by what one resident called "moose-infested forests" and located five hundred miles from the nearest large metropolitan area? How did this project's proponents position citizens in relation to this technology-led vision? Who had framed the project and with what consequences?

Municipal leaders described being inspired by Thomas Friedman's arguments about information technology building a "flat world" of equal opportunities within globalization. For three hundred years, Fredericton's main employers were the provincial government and its two universities and the local military base. University graduates could expect to walk down the hill from the campus to work in offices in the government buildings by the river. A prosperous community developed, with average incomes one-third higher than in large Canadian cities. In the late 1990s the provincial government eliminated many jobs, and the local government explored how to prevent a collapse of its economic base. In 2000 it developed an economic development strategy that focused on knowledge work and knowledge industries.

The mayor at the time explained, "We didn't want to have to be dependent on government, or so dependent on universities, which had served us well, but we wanted to diversify, and we had decided that information technology was the way to go." Drawing from the tight connections between local government and business owners who wanted to shift the economic base away from reliance on universities and government, this cross-cutting strategy focused on branding Fredericton as an innovative "knowledge-based community" to distinguish it from other cities in the region, all of whom were competing to retain young workers, increase immigration, and rebuild economies altered by the collapse of farming, fishing, or government-employment-based economies. Delivering on the promise of being a knowledge-based community was understood to require internet connectivity. Internet bandwidth in the region was at least twice as expensive as in major cities; some businesses had been paying eight hundred dollars a month for dedicated broadband lines. In response, the city government and several businesses investigated building their own co-owned broadband access network. Although this would create the backhaul infrastructure to deliver internet to businesses, it would not provide public access to the internet.

In 2004 the city government applied for a government-led Smart Communities fund to pilot a free Wi-Fi network. When the application was rejected, Cisco agreed to support the project through its Connected Communities project, providing Wi-Fi antennas and access to publicity and networking opportunities, including trips to Toronto and New York for the city's leaders. The city government linked wireless routers and antennas to their existing fiber backhaul, providing wireless internet access at no cost in public areas. This project was one of the first municipal projects to establish no-cost wireless internet access zones in airports, public buildings, and shopping malls, and it was virtually the only one to provision this no-cost access from publicly owned infrastructure.

The local decision-makers I spoke to described this project as a way to improve the business environment in their community. When I conducted fieldwork in the city government, project leaders mentioned that the businesses needing networks that had located in Fredericton (including Red Hat Linux and the FSC, which provided verification marks for forest products) would be able to offer their clients free Wi-Fi in their offices. Wi-Fi equipment was first mounted wherever there was available municipal real estate and, as the designers explained to me, was designed to work best in the center of town because "there are more business people there."

The positioning of the smart city as a facilitator of access, and access as an accelerator of business success, also shaped how decisions were made about where to locate access infrastructure. Downtown Fredericton is separated from a lower-income neighborhood and a First Nations reserve by a wide river. The access points supported by the Connected Communities project concentrated on the downtown side of the river, with a map of the project leaving out the opposite side of the river, which my interviewees characterized as rarely visited by tourists or business people. Nor was the Wi-Fi network intended to provide wireless access at any of the existing community technology access points, several of which were already in place in lower-income neighborhoods. One of the project's supporters said: "You end up wearing the problem. Who supports these sites, who maintains them, who puts registered software on them?"

The early proponents of the Fredericton network did not want to define their responsibility for public access to the internet in libraries or community centers. However, they did encourage media coverage that stressed the novelty and delight of accessing the internet for free, and the value of the novelty for the "brand" of local innovation. Consider the story I was told of a family traveling cross-country in search of a new home who parked in a Walmart parking lot and, after finding access to the internet so easy, decided to stay in Fredericton. In this early 2000s version of urban smartness, access to email on the go was almost magical, and connectivity across a city could be offered as a family's a reason to settle. In reality, immediately following the much-publicized launch of the wireless network, the city's chief technologist revealed that no rules were in place to guarantee how well the network worked.

My time in Fredericton occurred just after the opening of the network and its promotion in the international business press, but at the time most people used the network only occasionally, although students living in the downtown area used signal amplifiers to connect inside their houses to avoid paying for internet access.[13]

Although the Fredericton project consciously mirrored the frames set by technology companies and neoliberal government policy framing citizens as consumers empowered by "choice" to purchase communications access, it also hinted at another perspective on the nature and provision of communications access. In Fredericton, internet access was organized as a common resource available to all: the Wi-Fi installation, while marketed as a high-tech business amenity, was also a free service running on sponsored equipment and using excess capacity from a broadband infrastructure built as public infrastructure. At the time the network was built, the city's leaders were as reflective about how their broadband network acted as public infrastructure as they were about its potential to drive business creation. The director of corporate services at the time remarked, "Look, we provide tennis courts, all other kinds of infrastructure, so this makes sense. At the same time as we were doing this the Team Fredericton infrastructure was also being developed. It's a way of distinguishing us from all the other little cities in the middle of nowhere. . . . If you had the opportunity to do this, why wouldn't you?"

From the perspective of the network designers the project acted as "another layer on top of technology and innovation contributing to quality of life." This quotation from one of the city councilors I met explicitly defines how the city's decision-makers positioned it (like so many other cities, large and small) as "smart" and how strategies for city branding connected with the purchase of IT equipment more often than did inclusion strategies for marginalized people: "We are now selling the smart city. It is an economic development tool. . . . Wi-Fi is becoming the order of the day. It is like having better sidewalks—if you ask a city if they would rather have bad sidewalks or better sidewalks, they will always say they want better sidewalks. The overall principle doesn't change."

In Fredericton in 2007, citizens (or perhaps taxpayers) were framed as users of services—walkers on sidewalks, consumers of water, people accessing an information network. "Smart" citizens in this particular historical vision were anticipated as potentially able to collectively benefit from a shared infrastructure, but this collective benefit was always framed and justified in relation to business and economic expansion. Some citizens, if they were visiting the central business district, searching for real estate, or working in offices, played the role of ideal consumers of free Wi-Fi, while those needing access in libraries or community centers raised concerns that they might expect connectivity to be more than a luxury. In this vision of a neoliberal flat world of urban consumer citizenship the stress was on the economic benefit of expanded access to information.

OTHER VISIONS: WIRELESS CONNECTIVITY
AS A COMMUNICATION COMMONS

If the apparent magic of connectivity and the link between access to the internet and economic development underpinned how extending access became "smart" in Fredericton, other people responded with more radical perspectives to the possibilities of wireless technologies. Community wireless networks (CWNs), based on local experimentation with wireless radio technology, emerged around the world in the years following the drop in price of radio communication equipment that used unlicensed or license-exempt radio, which could be reconfigured using free and open-source software (F/OSS)—including the Cisco routers mentioned previously. These projects illustrated different ways of claiming communicative citizenship in different city spaces through "peer-to-peer" architectural processes aimed at creating communication commons.[14]

In 2002, the first "free information advocates" met in Berlin to talk about free information infrastructures—spaces where communities could self-provision access to information, including by building their own computer networks. These activists drew from principles of free software, reasoning that by using open-source licenses for the code that ran their network, they could create and maintain tools that would continue to be accessible. Many free software licenses use a viral principle of accessibility: specifying that software built using open-source code must itself remain open to future reuse. This self-perpetuating accessibility of software programs models an expanding communication commons supported by labor provided within developer communities.[15] The group in Berlin included both community organizers and people who had worked with free software; together, they thought that they might be able to use radio technology to extend access to the internet in a way that would self-perpetuate nonhierarchically. Using a radio technology called mesh networking that generated stronger links the more radios were added to the network, they sought to create a local communication network that would proliferate and self-perpetuate. An added benefit of a mesh network would be that if one radio on the network linked back to the internet, all of the linked sites would have access—making it possible to extend a single internet access point to a whole neighborhood.

In 2003, Freifunk, the first experiments with mesh networks at a neighborhood scale, started in Friedrichshain, in what had been East Berlin. At the time, this neighborhood, which had been cut off from more prosperous West Berlin during the Cold War, still had many buildings occupied by former squatters, and poor broadband access to boot. As Berlin rebuilt, many associations of squatters received rights of residence in their buildings provided that they continued to maintain them. Many formerly squatted-in buildings emerged into collective ownership and some, like K9, where one of Freifunk's founders lived, embraced radical principles of common governance. These principles included economic agreements limiting rents to the costs of maintenance and social agreements specifying collective decision-making supported by monthly meetings. These buildings could have dozens of apartments, and at K9, apartments were inhabited in rotation so that all building residents would have an equal chance to live in the larger or more beautiful spaces. This principle of fairness and justice resulted in a moving day every three years, when groups of apartment mates reconfigured as everyone moved house. I regularly visited with Freifunk activists between 2006 and 2012, checking in on how their radical visions for participation by providing connectivity complemented these other forms of collective social participation.

Like free software, the mesh radio networks that Freifunk used needed not only maintenance but also knowledge about how to keep the radio gear working so that people away from the internet access link would still be connected: in practice, this involved climbing up to the rooftops of the seven-story blocks of the neighborhood, checking cables, and sometimes writing (free) software to solve a problem. Having to work together and to create rules for things like ensuring that everyone on the network could use internet bandwidth (since one downloader might slow down speeds for the whole mesh) became as important as the connectivity provided. This principle of fairness and justice needed some work from everyone, much the way the mass moving day at K9 did. Participation wasn't equally distributed, though. In 2006 and 2007, the Freifunk network was large, but only about 10 percent of participants were doing what Juergen Neumann, an avid free software proponent and one of the instigators of the network, described as "social engineering." Juergen didn't mean social manipulation but social work as a type of connection or connectivity.

Perhaps not surprising for a project begun by people fascinated by the social models suggested by specific ways of writing software, the Freifunk project modeled participation in ways that connected to technically novel modes of governance. Armin Medosch, an artist, activist, and scholar of mesh networks, described the preconditions and consequences of participation in both the interpersonal and the technical links that are required to keep the network functioning: the advocacy for and use of a license-free radio spectrum. Interpersonal work is essential for making these decentralized networks operate. Because anyone who hosts an antenna is a member of the network, its function thus depends on creating and sustaining relationships, sharing knowledge about how to keep equipment running, and recruiting new people to join the network so that it becomes more robust as it expands. Medosch writes, "Self-organization is conceived as an active process, whereby economically and legally unencumbered participants voluntarily enter into collaborative relationships. This active, willful expenditure of personal energy, time and labor is made on the basis of joint striving to achieve a larger whole that is more than the sum of its parts: the network commons."[16] The network commons also needs to be maintained and defended by using the license-free radio spectrum. This part of the radio spectrum, outside the frequencies assigned to commercial or military use, is a diminishing commons itself, and activists explicitly used it to highlight how the airwaves could be of use and benefit to all outside of commercial licensing arrangements.

This model of communicative citizenship—or distributed smart citizenship—links together strong demands for interpersonal cooperation with claims and assumptions about the capacity of technology to mirror, or even transform, opportunities for people to participate. These first experiments in claiming communicative commons were followed by many others. Over the decade of the 2000s, hundreds or perhaps thousands of ad hoc or community wireless networks were established, bringing together people interested in experimenting with open wireless technologies and those interested in improving civic life. Not all of these projects used mesh networking technology or sought to explicitly create a communications commons by using a license-free radio spectrum. In the aggregate, they created alternative ways of claiming citizenship in relation to urban space—often in locations where relationships with the state were already contested.

The last story in this chapter is a more ambivalent one, where the discourse of the commons and the magic of connectivity combined into a well-intentioned but less explicitly radical proposal for community-based wireless access provision. In the mid-2000s context of heightened connectivity as a virtue in its own right, projects like Montreal's Île Sans Fil (Wireless Island) drew on claims to citizenship in relation to a local community or the digital divide to invite citizens to participate in creating their commons.

THE WIRELESS ISLAND OF MONTREAL — A TECHNO-COMMUNITARIAN VISION OF EXPANDED ACCESS

From 2004 to 2008 I was involved in a community wireless networking project in Montreal created by two young men with an explicit intention of "hacking the city" by inviting other young people to build internet access for use in urban public spaces like cafés, parks, and community centers. This project, called Île Sans Fil (ISF, named to reflect Montreal's location on an island), leveraged the enticing vision of the smart city as a connected space, as well as the excitement of being, as a participating citizen, a kind of hacker, critiquing and reprogramming urban systems.

Montreal has always had a strong sense of community and a tradition of self-organization, in part perhaps because of its status as the largest Francophone city in North America and its history as a city that has welcomed Jewish, Italian, and African migrant communities and has been a site of workers' solidarity movements, like the Montreal General Strike in 1920. In this context it is interesting to consider the backgrounds and expressed motivations of some of the original members of ISF. In 2004, Michael Lenczner, who had just returned from a government-sponsored trip to Burkina Faso to build internet infrastructure there, met with his friend David Vincelli to see if they could find a way to bring Wi-Fi internet access to the entire island of Montreal. They recruited entrepreneur Daniel Drouet and open-source advocate Daniel Lemay—and also actively looked for researchers to study their efforts, which is where I came in.

When I met "the Daniels" and Michael in late summer 2004, they had assembled a large group of young volunteers for the project who met weekly at a community-run pub in a somewhat seedy location near the downtown campus of the Université du Québec à Montréal (UQÀM), a French-speaking public university. With pitchers of beer on the table, the group switched between French and English, excitedly discussing plans to buy Wi-Fi routers, reprogram ("flash") them with open-source software, then distribute them to local businesses and community organizations, promising free support in exchange for an agreement that the organizations would share internet access with people who used their space. The offer from the Île Sans Fil volunteers was to provide the routers—the specialized software that collects information about who logs in at which hotspot—technical assistance, and branding, including a Google-map overlay showing the hotspot locations.

Between 2004 and 2007, ISF created a network of over 150 Wi-Fi hotspots. Through partnership with arts organizations they were also able to repurpose the software that kept track of who had logged in at which hotspot to stage creative projects: a short story told from various perspectives that could only be fully read by logging in at different hotspots, a prototype location-based social network that listed who was online at each hotspot, and a localized political news generator that identified the candidates for elections in the location of the Wi-Fi hotspot and presented this on the log-in screen for everyone who used it.

The creative hacker spirit evoked by the founders of the group sustained volunteer participation in the project for many years. New members were often recruited among students at the five universities and technical colleges in Montreal. In addition, through partnership with researchers like me and the consistent lobbying of journalists and politicians, ISF was able to position its volunteer-led, collaborative strategy for providing Wi-Fi within the broader tradition of community media, extending the idea of participation in media production into participation in creating and maintaining open-source software. My published papers joined celebratory pieces in *Le Devoir* and the *Montreal Gazette*, the papers of record in Montreal, which highlighted the community features of public internet access provision. By 2007, the long-term relationships had solidified public Wi-Fi provision as part of a new smart-city strategy, shepherded in part by Daniel Lemay, who took a position as an open-source software advocate within the city government in 2006. In late 2007 the Economic Development Commission of greater Montreal proposed a partnership with ISF to fund the expansion of the network to 400 hotspots, including 150 on city property. This agreement required the creation of a more formal organization, including a full-time, paid manager. By the end of the 2010s this project had become a municipally funded, arm's-length organization called Zone d'Accès Publique (Public Access Zone), which maintains the original hotspots supported by local businesses but also provides wireless connectivity inside major hospital developments.

The accomplishments of a group of volunteers in establishing and maintaining a large network of Wi-Fi access points is certainly laudable, but in retrospect the project raises interesting questions about the relationship between different kinds of participation in a smart city focused on access provision. To what extent did the tech-savvy, enthusiastic volunteers transform how technology was instituted or used? The members of Île Sans Fil held a sense of themselves as active citizens that contrasted with the way that people using the Wi-Fi network positioned themselves as passive consumers. The volunteers were comfortable with building and maintaining software and installing hardware—these were collaborative activities but not ones that were well understood outside the group. One volunteer described ISF as "primarily a social club for geeks . . . a club of passionate workers."[17] When I interviewed them, volunteers said that they wanted to be part of the project in order to contribute to their community. Many meetings finished with members introducing themselves and chatting, saying things like "we are really a nice bunch of people—we are the good ones."[18] The group remained relatively young, predominantly white, and mostly, though not entirely, male. Connectivity technology was understood as something that could actively bring people together. But who was coming together? One ISF group member wrote on the group's mailing list, "I'm very happy at how Wireless internet has taken me away from my indoor computer to the outside world. Today I meet many people, discuss how this technology can help communities, develop new potentials for people."[19]

These statements and actions suggest that the group had created an important public among itself, a kind of "geek-public." However, Michael Warner, a scholar of publics, argues that a public must continually extend its discourse to "indefinite strangers" if it is to be sustained: otherwise, the would-be public remains a closed group. Even though ISF appealed to indefinite strangers through its art projects and invitations to create early versions of social media profiles, the members who attended its meetings were more concerned in their meetings and discussions about "wiring up" locations than engaging indefinite strangers.[20] Daniel Lemay reflected: "It's as if we reproduced a production line" for the deployment of Wi-Fi hotspots. But the problem in reproducing an "industrial model . . . was that the people with the artistic projects were always outsiders."[21]

Observations and interviews conducted in November 2005 and May 2007 with people using ISF hotspots confirm the separation between the geek-public and the people using the network, as well as the way that the promise of connectivity acquired a magical dimension. The people logging on were demographically similar to the geeks: according to a survey of fifty-two people, around half were age twenty-five to thirty-four, and two-thirds had at least a bachelor's degree. Sixty-eight percent said that they used Wi-Fi hotspots "to get out of my home or office," suggesting that public access did have some features of conviviality and community building. However, people using the network said they were not primarily motivated by community or conviviality but more by not paying for connectivity—one interviewee described himself as "opportunistic—but aren't we all?" In addition, many of the people I interviewed preferred accessing Wi-Fi networks anonymously and were annoyed with ISF's authentication procedures. The fact that the service was free—as in free of charge—was considered more important than the fact that ISF's technical and social structure were open to participation. People I interviewed knew that ISF was a community organization, yet none of them had attended meetings, although one volunteer said that he had "given them [ISF] my opinion on a couple of things, but they always ignored me."[22] As the project matured, the openness to participation declined, and the geek-public became more distinct from the community-public. While the geek-public could maintain its convivial connections by meeting to talk about and build Wi-Fi networks, the community-public never shared the same space of public engagement. The professionalization of community wireless activism in Montreal enhanced the focus on the labor and influence of individual geeks—especially as recruitment for new volunteers turned increasingly toward students at a local university. Participating in the project became more akin to undertaking an internship, as the novelty of hacking the city in this way declined. Nevertheless, public connectivity provided in a nonmarket manner and community participation in defining the benefits of technology remain aspects of Montreal's official smart-city plan, in contrast to that of many others.

COMMUNITY-PUBLICS, GEEK-PUBLICS, AND THE COMMONS

These stories show how the promise of technology influenced the ways that citizens were invited to participate in early-2000s versions of the smart city. The legacies of these projects illustrate the influence of assumptions about the technical capacity of networking, as well as its economic or social significance in either bringing investment or bringing people together. The city government in Fredericton succeeded in creating common, publicly supported infrastructure that could compete with commercially owned infrastructure, even when the descriptions of and justifications for the public Wi-Fi network consistently referred to economic competitiveness and business success rather than communication access as a right to be exercised by all. Advocates in Berlin built a mesh network to provide access when state and market failed, but their model of governance through physical network maintenance was too burdensome to maintain as commercial access became more available.

A decade and a half after the Freifunk activists refused to declare themselves an organization or propose explicit organizational strategies, Freifunk organizations exist in most German cities. A cofounder of Freifunk expressed some dismay to me recently that one of the volunteers considers that he is "in charge" of the local network and that others see volunteering as a good way of gaining skills in order to join mainstream telecom companies—exactly the kinds of organizations that freenetworking proponents hoped to undermine with their proposals for horizontal organizing and self-management. Equally, while Freifunk is active across Germany as a social group for geeks and a place for young people to learn about networking and meet others with similar interests (like ISF's successor has become in Montreal), and the organization has also provided internet access in public locations and in a refugee camp, the maintenance and expansion of the autonomous network in Berlin have become difficult, since not as many people want to be responsible for tinkering with antennas in order to take responsibility for a set of social relationships. Communication scholar Andrew Herman explains how so many technology activists evoking commons or collaboration using technical means are disappointed when the social outcomes do not easily follow: "The commons paradigm tends to assume in reductionist fashion that the correct technical and juridical disposition of people in relation to things will, *ipso facto*, yield a communicative relationship of communality where people anywhere can connect to each other everywhere based upon an ethical sharing of communicative resources." He further notes that "the commons as sphere of communication, the commons as a place of community, and the commons as a property relationship regarding technical resources become elided." Medosch makes a similar point: that the network commons discourse is separate from the discourse on economies of solidarity that would facilitate an economic transformation to the commons, and separate also from forms of radical collective engagement and mutual aid.[23] While both Fredericton and Montreal developed public organizations that, in different ways, supported continued access, the organizations also reiterated the connection between access to the internet and economic progress (in the case of Fredericton) or community engagement (in the case of Montreal's Île Sans Fil). The self-consciously radical ethos of Freifunk paradoxically allowed the model to proliferate across many cities, although this came at the cost of a disconnect between anarchic groups and their strong communitarian roots.

NEW DIRECTIONS FOR ACCESS TO TECHNOLOGY AND COMMUNICATION RIGHTS

The lessons from ISF in Montreal include the observation that technical expertise and enthusiasm for new technology separated out two publics who might have been unified through a different understanding of community technology.

Drawing on the identification of the gaps between the geek-public of advocates like ISF and the potential for broader community participation, democratic technology researcher Greta Byrum has investigated ways to explicitly connect community-based technology provision to social goals, reinvigorating the idea of the communicative commons. She identifies how the Allied Media Project based in Detroit facilitated the development of local knowledge and social capacity by adopting models of distributed mesh networks. She writes, "The Detroit Future Media (DFM) program, led by Diana Nucera, created a 'Digital Stewardship' training curriculum. . . . Nucera built the curriculum using a popular education method grounded in the history of the Civil Rights–era Citizenship Schools and Paulo Freire's *Pedagogy of the Oppressed*. She created this people-guided approach to bring tech education to communities that have been harmed and oppressed by technology, such as those in Detroit, where factory automation has killed so many vital middle-class jobs."[24]

In this adoption of mesh networking technology and participatory strategy, community capacity building has come first. The ideas behind the DFM project have also influenced community organizers based in Red Hook, Brooklyn, who built an autonomous mesh network to reflect the existing social networks of care and solidarity. Through further related work, Byrum and her many collaborators have created sets of portable networking kits that can work in an autonomous or distributed manner. These portable networking kits (PNKs) bring a punk DIY approach to community-led technology work, focusing on technical work as a type of care and maintenance, similar to that of the Freifunk project. However, by bringing the related goals of local autonomy and community-based access provision to the deregulated, liberalized context of the United States, these projects show a global continuity to commons-based ideas of communication rights.

The examples in this chapter illustrate that the smart-city vision based on connectivity and access to shared information infrastructure can create different ways of approaching the creation, maintenance, and governance of the infrastructure underpinning them. Providing connectivity promised a possibility for linking the space of flows with the space of places and redefining these connections in ways that supported a vision of the commons. These commons visions, like the ideas that promised transcendence through the embedding of global IT in cities, sometimes overstated the connection between the organizational capacity of the technology and the economic or cultural practices underlying them. In Fredericton, the local government responded to the invitation from technology companies to connect the city by defining connectivity as common infrastructure to be maintained for the use and benefit of all. Maintaining and upgrading this public infrastructure would then be the responsibility of the local government, acting to secure infrastructure as a means to ensure access for all. In Berlin the absence of a functioning support for developing and maintaining infrastructure created a space for activists to imagine different ways of providing connectivity and to give themselves a role in defining how it might be done. The radical claims made about the relationship between the creation and maintenance of this distributed, non-hierarchical network and community self-organization are generative as well as seductive. Reorganizing provision of network infrastructure based on principles of mutual aid and self-sufficiency demonstrates another direction away from commercial monopoly. However, focusing too much on the requirements for technical knowledge as a contribution to maintenance took the Berlin mesh network project away from its community-supporting roots.

As activists learn from these early experiments and reposition connectivity technology against a backdrop of community resilience in the face of economic instability and climate change, the notion of the commons reappears, this time challenging a state of urban smartness that is increasingly tied up with the possibility for pervasive connectivity to facilitate pervasive data collection and "surveillance capitalism."[25]

CHAPTER TWO

Data Cities and Visions of Optimization

I N the past decade, the dominant vision for the smart city has shifted from a communicative city equipped with access to an information society and toward a pervasively connected city where governments and commercial enterprises can use data to increase efficiency. I call this vision the "big data optimized city." Connectivity is no longer the goal; now it is a precondition for the intensification of data collection. In turn, data can be processed to generate efficiencies—to optimize transport, waste disposal, and other service delivery or to remake forms of citizen participation like building apps or collecting data. Like access, optimization is a vision that transforms the arrangement of technological and social resources in a city, laying down tech-driven assumptions about how social life should unfold.

In this chapter I outline how a new framework for techno-systems decision-making about the smart city has emerged—a big data optimized city. This framework has emerged in places where connectivity and access have been secured; both of those features open new spaces for economic exploitation as well as new expectations for civic participation using data. We can see how the promise of the big data optimized city is framed through the language used by companies that control the self-learning feedback systems that keep cities running. We see how the same platform-based economic and organizational models that these companies use shape the way that citizen-based organizations work, too. This overall process of optimization narrows the frame for citizenship; individuals are perceived as consumers who can be nudged to change their behavior based on predictions extracted from data they share. As communication scholars José van Dijck, Thomas Poell, and Martijn de Waal argue, "The adoption of platforms causes a clash between stakeholders and public values."[1] Here I explore this clash and its consequences through the example of the emergence of the big data optimized city.

According to authors Viktor Mayer-Schönberger and Kenneth Cukier, datafication reduces human lives to records in digital data, which can then be aggregated and analyzed in ways that produce value.[2] In putting this perspective into practice, big data optimized cities depend on intermediaries—organizations that can deal with complexity and create stories from data. These computational intermediaries process data and issue predictions, insights, and patterns to governments and other decision-makers. The work is significant, because it influences operational decision-making and policy-making in cities. Data-based intermediaries gain more significance in the big data optimized city because the smart-city model has moved from the network to the platform. Urban platforms can include dashboards that present data collected by sensors, route-planning systems that make predictions for drivers based on data collected by GPS, and urban navigation systems that combine data from transport services and drivers. These intermediaries, and their use of calculation and optimization, create frameworks for communication and citizenship that fit an "optimization frame" that pre-assumes connectivity and absorbs data.[3]

For example, electricity consumption data aggregated by smart meter systems brings into being an "electrical data subject" with a lifestyle observable through its data traces. So the data from my apartment in London produces a new Powell Family Data Subject composed of the records of electricity used at any time—when we turn on the lights, run the washing machine, open the laptop, or turn the heat on when the weather gets cold—creating a proxy model for the dynamics of family life that includes how many people are resident in that house at any time.

In other words, urban smart technologies proscribe some actions and facilitate others—which can be as persuasive as the rhetoric about the best way to use these systems, their inherent value, or their ability to improve the world in very particular ways. The same smart meter that transforms my family into a new kind of data subject makes it possible for the electricity company to manage demand for electricity very specifically, and for me to know exactly when it is cheapest to run the washing machine. However, it is much easier for the electricity company to extract information from the hundreds of thousands of smart meters that it operates than it is for me to understand mine: that information asymmetry again.

The smart meter is a platform that allows the electric company to do various kinds of intermediation based on data (management of demand, delivery of targeted upgrades, other kinds of optimization). Other kinds of platforms, such as social media platforms, also collect data on individuals and, using the collected data, are able to segment audiences, conduct specific profiling, and target advertising, all without having to spend money to create content. A shift toward the platform has promised to make cities into cybernetic systems, self-optimizing and oriented around data and predicated on information asymmetry.

DATAFICATION AND THE CYBERNETIC CITY

To understand the implications of this shift from the communicative to the data-optimized smart city, it is important to see the relationship between datafication and network access. Underpinning claims for economic rationality of datafication lies a cultural assumption that the world can best be understood through data. Communication theorist José van Dijck writes, "Datafication as a legitimate means to *access, understand* and *monitor* people's behavior is becoming a leading principle, not just amongst techno-adepts, but also amongst scholars who see datafication as a revolutionary research opportunity to investigate human conduct."[4] Urban datafication depends on city residents to (willingly or not) produce data intended to optimize city services. Under governance regimes that include budget cuts and diminished responsibilities for public institutions, urban citizens are repositioned as consumers of city services and producers of the data that make platform models profitable. These consumer-citizenship positions support a vision of a city as an integrated service-delivery platform, with services and processes streamlined. As governments respond to promises of cost-cutting and reduction of employment, swaths of services, including the delivery of social support benefits, are delivered via automated systems.⁵ As these roll out and as poor people are increasingly judged by their data traces and enrolled into systems that don't account for their situations, it is helpful to see how these systems became so attractive and what the consequences are of a focus on optimization as the new frame for urban relationships.

The big data optimized city is different from the access-focused smart city. In the paradigm of communicative citizenship, the right to communicate is positioned as a right to gain access to a network, which is interpreted as if it were part of an expanded capacity to communicate. As we already saw, this interpretation foregrounds economic arguments about the value of connectivity. Once that connectivity is in place, it facilitates a data paradigm that assumes that important aspects of life can be rendered into data and, once made into data, can provide insight that cuts costs, optimizes processes, and renders urban life more seamless.

In communicative terms, datafication shifts the focus from content to form. As Louise Amoore, scholar of algorithms, writes, "The allure of unstructured data is that it is thought to contain patterns heretofore unseen and, therefore, a wealth of previously hidden insight. The growing use of analytics capable of reading and making sense of data, of unlocking its potential, is tightly interwoven with a 'world of promise and opportunity' thought to be buried in a text in need of an index." More specifically, in a framework of datafication the act of communication comes to take more significance than the message it is meant to carry, and the expressive capacities of citizenship start to be associated with data analysis rather than expression, becoming, as Evelyn Ruppert and colleagues write, "a politics of mash-ups, compilation and assemblage."[6] Datafication also contributes to the valorization of norms of publicity—information, communication, and participation—rather than the political ends that these are supposed to serve. It produces communicative capitalism, to again use Jodi Dean's descriptive phrase. In other words, the act of communicating a message has come to stand in for the message itself. The message is part of a data stream, and its most important feature is its circulation, which is managed and mediated by companies and organizations. The constant flow of communication allows data to be brokered, financialized, and made valuable in many of the same ways that credit and financial information became valuable.[7]

CYBERNETIC CONTROL AND DATA ASSEMBLAGES

Under regimes of datafication urban residents are encouraged to become consumers of the services they optimize by sharing their data. This creates a cybernetic feedback loop where the functioning of (commodified) service systems needs to be sustained by contributions by their beneficiaries. Cybernetic proposals for management of complex systems like cities have been around for decades. They suggest systemic analysis and decision-making as ways to render cities "legible." Many proposals for data-dependent smart cities leverage visions of cities-as-systems that draw on histories of cybernetic control, which refers to self-learning systems. Automated feedback mechanisms and the central control of data assemblages have been compelling visions since theorist

Stanford Beer's proposals for a networked cybernetic system for the Allende government in Chile in the 1960s.[8] New visions of cybernetic control are now made concrete in projects like city dashboards that combine real-time urban data indicators to "visualize" city functions like public transit, road congestion, crime statistics, and weather information. Geographer Rob Kitchin and his coauthors write, "For their advocates, the power of indicators, benchmarking and dashboards is that they reveal in detail and very clearly the state of play of cities. They enable one to know the city *as it actually is* through objective, trustworthy, factual data that can be statistically analyzed and visualized to reveal patterns and trends and to assess how it is performing vis-à-vis other places." The city becomes visible and manageable as a set of indicators presented on a single interface, abstracted out of time and space.[9]

These renderings of the city into data and representations of indicators are models, not reality. The data assemblages actually shift and develop all the time: "the apparatuses and their elements frame the nature, operation, and work of an indicator initiative. And as new ideas and knowledges emerge, technologies are invented, skill sets develop, debates take place, and the form of city governance alters, the data assemblage evolves and mutates." Regardless, this cybernetic history of the optimized big data city is visible in urban "control rooms" where streams of data from many different sites are aggregated. In control rooms, the promise of optimization is visibly rendered in large screens showing real-time indicators. The smartness of these big data optimization systems accrues through the seamless, constant calculation of data that are then presented back through dashboards and screens. In this process there is a constant attention to the future, with systems aiming toward optimization of services and reproduction of the calculative machinery—aiming to continue to accumulate ever more data in order to improve the algorithms that underpin these smart systems. As Orit Halpern writes, "'Smartness' becomes a new catchphrase for an emerging form of technical rationality that, by continuously collecting data in a self-referential manner, is able to constantly defer future results or evaluation and assume the responsibility of managing this structurally uncertain future."[10] The cybernetic model learns from data, but feedback loops are usually imagined as closed, which produces a different set of relations from the generative, horizontal ones in the distributed mesh network.

From the citizen perspective, being a source of data for someone's dashboard is different than receiving access to the internet and being invited to use that connectivity to express oneself. The big data optimized smart-city vision places value on the generation of data and its calculation, curation, and aggregation, making an explicit assumption that these activities, rather than the generation of ideas or knowledge, are what matters.

The big data optimized city has depended on the efforts to extend connectivity. These efforts have involved a move away from considering the communicative potential of citizenship and the need for access to a network toward considering citizenship as producing data for action. This shift has big implications for people and for institutions, because it changes the kinds of intermediaries at work. Rather than being organizations providing access, these intermediaries collect, process, and present data. Just as companies like Cisco and IBM once created strategies to encourage citizens to participate in expanding access to networks, big companies (sometimes the same ones), as well as governments and third-sector organizations, are now developing ways to benefit from or intervene in data collection by using the promise of algorithmic insight and the architecture of the platform. The real influence of these models, however, is that they also shift how active citizens are expected to engage with their cities. Datafication, intermediation, and platformization act as affordances for civic action: ways of shaping and influencing the capacity for social and cultural action to take place.

DATA-BASED VALUE AND OPTIMIZATION

In the data-based smart city, datafication, intermediation, and platformization are framed and put into practice by a set of commercial actors whose interest in defining and controlling data systems leads them to mobilize these ideas in particular ways. Civic actors, too, must contend with how data produce value, shape optimization processes, and fit into cybernetic feedback loops directed at reducing risk and optimizing prediction.

In order to produce value from a set of data, an intermediary will use algorithmic techniques, defined as "encoded procedures for transforming input data into a desired output, based on specified calculations."[11] The cybernetic systems that underpin the framework of the data-optimized smart city are intended to reduce complexity in various aspects of city life. For example, smartphone apps can suggest straightforward walking routes to unknown destinations, make it easy to find restaurants similar to those you have already eaten at, or predict accurately when the bus will arrive. Of course, the logic of predictability and, indeed, possibility have deeper social consequences—because the big data produced in smart cities requires "small analytics" that, by aggregating, parsing, cleaning, and ordering data, enact particular social assumptions.[12] This cleaning, ordering, and parsing is needed to make a cybernetic feedback system return optimal results, but the process removes most of the traces of friction and difference that are part of the reality of urban life. Friction is removed in part by justified efforts toward optimization and in part by the construction of data-based optimizing systems, which are based on the model of a multisided platform that emerged from the technology industry.[13]

The data-based optimized city is predicated on the idea that a city can work like a platform: a computational arrangement that permits exchange and access of digital resources based on common standards and that links together two or more entities through a single platform intermediary. Technology researcher Tarleton Gillespie identifies four "semantic territories" for platforms: architectural, figurative, political, and computational.[14] Architectural platforms include oil platforms and station platforms, while a figurative platform is a metaphysical base from which further opportunities can be found, and political platforms put forth politicians' views and visions. Following from van Dijck and Poell's insight that the discourse and organizations of platforms are performative, we can track how discourses and organizational models for platforms and optimization flow from technology companies through governments and to civic organizations.[15]

The concept of "platform government" was developed by digital advocate Tim O'Reilly in 2011.[16] Throughout the 2000s, the idea of rendering government into a form that would expand openness, participation, and efficiency in service delivery inspired efforts to model smart cities as platforms—and citizens as data sources.

The National Endowment for Science, Technology and the Arts (NESTA), the UK innovation charity, argues that "platforms are about providing a (digital) framework within which others abide by rules, using data and a payment and regulatory ecosystem to unleash invention at scale." What this has meant in practice is the creation of straightforward processes for data processing, validation, and payment for services. Yet the consequences of this enthusiasm for the platform model have also brought changes to how urban governance is understood. Governments manage data and provide a thin layer of rule enforcement of the brokerage provided by third-party operators. As information theorist Ellen Balka describes it, information systems produce "shadow bodies" by emphasizing some aspects of their subjects and overlooking others. As Gillespie puts it, "these shadow bodies persist and proliferate" as "either politically problematic, or politically productive."[17]

These shadow bodies produce data that become both a product and a material. As a product, it can be collated, bought, and sold across a platform. As a material, it can be analyzed, bought, sold, and brokered on a platform. Data can also be the basis for financial speculation by the brokers that hold it. With the expansion of algorithmic judgment, more and more parts of life become black-boxed, with access to credit, policing, or commodity pricing related to features surfaced in data analysis.[18] In urban life, expansion of data collection promises to reduce risk and increase rewards for data brokers. Optimization of city services, from the perspective of the government-as-a-platform, means reducing risk by delegating as much operational control as possible in return for lower costs or clearer metrics.

The big data optimized city, like the networked smart city, is based on interpretations of technological possibilities that align with the interests of companies and entities able to take advantage of them. As a "data imperative" spreads, companies and governments assemble the resources required to enact it; the resources include the systems for data-based management and the discursive justifications that support them. Corporate smart-city projects attempt to move all aspects of data collection and processing to corporate control while promising risk reduction for city governments.[19] This results in operational decision-making shifting away from the city government: an attractive proposition when local governments are obliged to cut their budgets. Instead of hiring a statistician on a salary, a government can funnel public data into a platform and gain access to analytics.

CORPORATE ACTORS AND THEIR PLATFORM STRATEGIES

Some of the actors in the big data optimized city are familiar. For example, after selling equipment promising broader connectivity via Wi-Fi for a decade, Cisco rebranded to market its equipment as the backbone of the "Internet of Everything." It created a platform to connect already-installed wireless technology in cities—but not one that provides citizens a capacity to be heard about issues, nor even one that provides individual citizens access to the internet or to sensors. Instead, Cisco's Smart + Connected Safety and Security product promises to link together the company's existing urban Wi-Fi provision with "IP cameras, sensors, and smartphone apps. . . . [It] integrates with enforcement applications, pushing expiration notices to traffic officers for ticketing. It also provides visibility into parking analytics, including usage and vacancy periods, which helps cities with long-term planning."[20]

This smart-city system transforms car movements into data and facilitates action that consists of enforcement of parking tickets. This happens by pulling data into the Cisco cloud, processing it, and pushing notifications of stationary cars into portable devices held by enforcement officers—or, in an even smarter city, issuing tickets electronically without recourse to human action or judgment. This increases the transparency of people's movements to the city governments and the parking enforcement companies but removes public oversight by parking attendants as well as the clear institutional chain of responsibility between the government, the enforcement officer, and the car. The product is also fairly banal, given that it is the pillar of Cisco's new Smart strategy. What it does do effectively is continue to lock governments into buying, maintaining, and relying on Cisco router technology—which is no longer sold as a way to provide access to the internet but instead as essential infrastructure that will underpin Cisco's marketing of analytics technology and cloud services.

Since smart-city intermediary companies benefit most when they succeed in becoming the main platform integrating all streams of data, each company seems to make promises more overblown than the last. Siemens, originally an electronics company, declared in its 2015 smart-city branding that "software—both standalone and embedded—is an integral part of almost every Siemens product. With our know-how, we can intelligently manage the masses of data generated along the entire infrastructure value chain, helping to interpret data correctly." The promise to encapsulate an entire "value chain" means integrating machine learning and other automated decision-making processes. The ability to "automate infrastructures" and "constantly refine itself" to make it possible to "one day respond to events as they occur" offers an alluring possibility of cybernetic control.[21]

For a platform company to collect data once and sell it twice is an effective business ploy. For example, commuting applications gather government open data on transit arrival times, as well as individual location data from people who use commuting or route-planning apps. With the data, transit authorities can plan the movement of people through transit systems, and individual commuters can optimize their own movement by following the app's instructions on paths to pursue or avoid.

Let's look at the trajectory of a startup company to see how urban data collection generates two products—one for city governments, who are promised a way to optimize urban travel at an abstract level, and one for individuals, who are promised an optimized way of moving around—and how the data behind these products made the company worth acquiring by Google (now Alphabet, the world's biggest platform company). In 2014 a startup company called Urban Engines invited commuters to have their commuting measured through a GPS-enabled application on their phone and in exchange to be notified of ways to optimize (i.e., reduce) their commuting time. The company's original description of the product said that it "provides rewards for small changes to commuting times that reduce congestion." Aggregate analytics based on the data of many commuters were then provided to institutional subscribers (city governments). As a bonus, the app could micro-target commuters to encourage specific types of behavior. This is not a neutral power relationship, even when the individual benefits. But Urban Engines also promised (like many other smart-city analytics companies) to use feedback models from computing to build a citywide operating system. In 2016, when they were acquired by Google, they described their contributions in this way.

> Earlier, we had worked on making the web faster, better and more personalized. We sensed an opportunity to apply those lessons to the world we live in, and make urban living better. With the rapid growth in sensors (GPS, beacons, etc.) on smartphones and cars and transit, we are now truly in the age of Internet of Moving Things. The potential to improve the lives of millions of commuters, by learning from commuting behavior patterns, reshaping congestion, and creating new consumer services in each minute-mile is incredible. In fact, we have an opportunity to create an urban OS—an intelligent software overlay for our real world.[22]

Beginning in 2016 this overlay was integrated into Google Maps, then as now the dominant digital city overlay. Google has succeeded in making its computational infrastructure essential for the operation of the smart city not by repurposing access technology like Cisco, nor by creating its own stand-alone platform like Siemens, but by operating as a platform and then integrating its services into similarly structured urban platforms.

The platform structure also provides opportunities for companies to sell services at the interface. Early smart-city systems sold services that identified and controlled risk and positioned any civic action outside of a certain set of actions as threatening. Like Urban Engines and other data-based brokers, in the mid-2010s dozens of companies promised to mitigate risk for urban governments. For example, DataSift, a brokerage company, declared that "whether you need historical or real-time data, feeds, or firehouses, one REST API is all you need for 100% licensed access to a deduced data stream. If a connection goes down, we handle it, not you." Recorded Future promises a consistent scan of cyberthreats: "We do the hard work for you by automatically collecting and organizing over 690,000 Web sources to identify actors, new vulnerabilities and emerging threat indicators. Gain deep visibility into your threat landscape by analyzing and visualizing cyber threats, even across foreign languages, with our vast open source intelligence (OSINT) repository."[23] These risk management strategies position data intermediation and commercial management of the cybernetic circuit as essential to mitigating the risks inevitably associated with managing cities as data streams. As platforms and data collection expand, this risk management is essential for maintaining control of systems. Contributions from individual citizens are welcome, but they are also seen as inherently risky. Collected data needs to thus increase in scope constantly, so that data that threaten the function of a feedback model can also be collected.

ANTICIPATORY MEANING-MAKING

The final element of a cybernetically optimized big data system is the ability to predict or make claims about future states based on the data gathered. Science and technology scholar Adrian Mackenzie identifies the prospective mode as being one of the key ways that machine learning is changing knowledge production. With machine learning, systems are constantly working to gather patterns and propose new directions.

Prospective modeling systems have been heavily marketed to cities, although the extent to which they are effective is debatable. For example, analytics intermediary Criteo promises that its new analytics engine "uses artificial intelligence and machine learning technology to automatically analyze data, detect patterns, build statistically validated predictive models, and send information to virtually any type of application or technology."[24]

The analysis underpinning the data presentation and especially the predictions that are the core value promoted by these companies depend on different forms of data abstraction. The use of sophisticated data management technologies like machine learning promises the ability to optimize experience by predicting it. The goal of such prospective systems is often to change the behavior of urban residents. Combined with "nudge" psychology, analytics permit measurement of individual behaviors and—importantly—analysis of relationships between individuals. For example, behavioral economist Andrew Meaney argues that applications like Citymapper and Urban Engines permit policy-makers to reframe different transport choices as socially desirable: they can "implement congestion mitigation policies that nudge individuals towards socially optimal travel decisions without restricting their options or changing financial incentives."[25] Nudging, especially for transport decisions, depends on the nudger having collected and processed data that permit them to model more or less optimal decisions for others. This modeling increasingly involves prediction based on automated judgment.

Platforms introduce particular ways of thinking about citizens and data. Machine-learning processes necessarily make assumptions about how social life is organized. Their mathematics depends on being able to assume that responses to situations can be generalized; further, the process of generalization works by "imposing some kind of shape on the data and using that shape to discover, decide, classify, rank, cluster, recommend, label or predict what is happening or what will happen." Mackenzie, as well as critical computation scholar Dan McQuillan, argues that these generalizations are made by employing mathematics that applies vector analysis to multidimensional data, which has also driven the development of new database systems that seek relationships among many different variables, rather than databases where queries or relationships needed to be structured.[26]

IMPLICATIONS FOR CITIZENSHIP

At the intersection of predictability and risk management are people, their everyday practices, and the data that result from them. If we take citizenship to be related to the capacity to communicate in relation to things that matter, and to employ aspects of the digital in relation to democratic aims as opposed to commercial ones, then citizenship in a big data optimized city remains constrained, since no engagements are invited or expected beyond communication of information in the form of data. Furthermore, the imposition of operating systems onto cities and the third-party commercial control of the data flowing in and out of these systems provide visibility and control of the systems only to a small number of operators.

The move from network access to cybernetic datafication shifts the kinds of promises made in relation to city function away from ideas of linked and networked relationships and toward ideas of cities becoming fully represented in data and made meaningful through processes of analytics. The logic of machine learning introduces mathematical specificities that dictate how experiences might be able to be generalized: "generalization depends heavily on specificity, including many domain- or algorithmic-specific details that rarely surface in the romantic employments of machine-learning-based prediction as generalizing patterns in the data."[27] Many of the leading applications for civic use assume that everyday life is oriented around consumption and that individual experiences of everyday life, especially mobility, can be optimized through management and mitigation of risk via consumption of new products. Indeed, applications like Urban Engines acquire real-time GPS data from anyone who uses the application (or now, from Google Maps navigation features). Once acquired, the data are used to improve the recommendation algorithms, increasing their precision. But a few obvious issues with this become clear: What about the movements of people who do not have smartphones? Buses can be tracked in these optimization processes, but pedestrians and cyclists may not be unless they carry smartphones. It is worth considering that the deeper logic of these systems is not really directed toward making transportation more efficient but actually toward creating detailed, proprietary data about the movement of people in space, which can be beneficial for training future machine-learning systems—most of which are likely to be used by future data intermediaries to produce future strategies for managing movement in cities.

On a conceptual level, machine-learning processes depend on using a set of very specific information to extrapolate multiple dimensions to produce generality. These processes reinforce governance processes oriented around predictability, but they also introduce a further feature to that predictability: machine-learning techniques require a certain amount of stability in order to generate predictions. As Mackenzie points out, mathematically, machine learning systems "struggle to predict becomings." Machine learning's production of generality and prediction might one day produce new forms of social relations, but in its current state the generalizations associated with its predictive power are employed to produce a predictable consumer-citizen. The predictable citizen's future actions, movements, and desires can be generalized from machine-learning technologies and fit with both the consumer-citizen and the operation of the risk society, in which collective experience is oriented toward risk, especially its mitigation. This description echoes the analysis by Amoore that decisions about the movement of people and the security of borders are also primarily based on risk calculations and the development of data derivatives that can "generate a good enough decision." The use of derivatives changes the nature of risk, making it prospective: "risk in its derivative form is not centered on what we are, nor even on what our data might say about us, but on what can be imagined and inferred about who we might be."[28]

The design and organization of many of these applications and platform systems suggest an ideal consumer-citizen who is not a liberal subject choosing among options but who instead is a data producer and consumer whose movements can become valuable material for an intermediary. This value accrues to the intermediaries who have the closest to total control (and visibility) of the analytic space. Urban Engines, for example, can model the outcome of traffic patterns using various assumptions and then use the models to define the algorithms that now suggest routes on Google Maps.

However, this process can only be understood, tinkered with, or challenged if someone is able to see all of it. A system that concentrates financial power in the hands of centralized platforms is what sociologists Paul Langley and Andrew Leyshon call "negarchical capitalism."[29] The city-as-platform formulation extends this financialization of citizenship by promising to deliver optimal experiences to citizens and optimal data flows to governments. But what it actually does is to narrow both citizenship and governance with the promise of relating them through a market. The eventual outcome of this marketization is never pure competition but monopolization. The concentration of power in a shrinking number of platform intermediaries illustrates how power proceeds from the ability to control the different sides of the multisided data market that intermediary participation creates within the city.

The rise of the platform city and the reliance on data-driven intermediaries gives structure to an urban governance partly concerned with the pursuit of optimization. Optimization has itself become a framework for citizenship—a model of consumer choice extended to the provision of services and the everyday experiences of people. Optimization depends on effective prediction—theorists of governance who follow Foucault have identified how technologies of rationalization have positioned certain kinds of civic acts as desirable and others as undesirable.[30] So how does optimization change with expectations of total cybernetic control, risk management, and prediction?

When a main framework for civic life is optimized, some things are going to be easier to fit in the framework than others. It is relatively straightforward to optimize transportation or the collection of recycling but more difficult to optimize volunteering, knowing neighbors, or creating local capacity to take care of people in a crisis. This difficulty raises some interesting issues of ethics that produce dilemmas of contemporary technological citizenship, as civic organizations appropriate the language and design practices that orient toward optimization. Datafication, intermediation, and platformization appear in civic strategies as well as corporate ones, suggesting that a techno-systemic frame oriented toward optimization has shaped ideas about active citizenship as well.

OPTIMIZABLE CITIZENSHIP — PUBLIC SECTOR
AND CIVIC ORGANIZATIONS

So far I have outlined how technology companies have defined their role in managing and brokering the data contributed by consumer-citizens. What is more significant is that the same ways of thinking about data and civic participation come from bottom-up, citizen-focused intermediaries. These groups appeal to the logic of optimization as embedded in the platform frame. FixMyStreet and CycleStreets, both UK projects, illustrate some of the earliest attempts to reappropriate the big data optimized city, and the limits of these attempts, as does SeeClickFix in the United States.[31]

The civic technology consultancy MySociety launched FixMyStreet, an interface for crowdsourced contributions of local problems, in 2008. This platform collects location data points and user-generated content identifying maintenance problems that cities should fix. It also allows governments to reduce their expenditure on maintenance by identifying areas of maintenance that, if addressed, will be positively viewed by the people who submitted data. This is meant to optimize—make more streamlined and efficient—the relationship between the government and these people. Unfortunately, this relationship cannot account for the views of the people who have not used the platform. On one level, these people might be excluded from access to communication networks, but on another level, their failure to submit data to a platform system that could calculate it into something that would make government's work optimal removes them from consideration. Critiques of Fix-MyStreet have pointed out the number of "languishing" reports that do not trigger action, as well as the distortion that crowd-sourced demands for government response can produce, where local governments fix potholes identified by affluent residents with smartphones and eliminate core funding for services such as libraries whose value is not as easily and immediately datafied.[32]

CycleStreets, a nonprofit organization that develops cycling maps based on contributions from individual cyclists, developed a trip-planning and problem-reporting application in Hackney, a London neighborhood with a very high level of cycling commuters, in 2016. The app collected data from the GPS function of cyclists' mobile phones and provided this, along with information on the purpose of the trip and basic demographics, to Hackney Council so that the council could feed these data into their transport planning, as well as use it to populate online maps and, for a time, its own dedicated app. CycleStreets was open source and built from the OpenStreet-Map open-source map database, which was itself created by mapping activists associated with the free-software, mesh-net-working, and free-information-infrastructure movements.

As of 2019, CycleStreets had produced its own UK-wide route-planning map for cyclists, as well as an online map package for cyclists that is built up from OpenStreetMap. In order to develop and maintain the app, the CycleStreets founders established a not-for-profit company, but the mapping systems continue to solicit support from cyclists willing to share their mapping data. CycleStreets remains a resource built by cyclists for cyclists, but because of its small scale and open-source approach it doesn't work as seamlessly as Google Maps. Paradoxically, by being one of the first route planners to solicit information from cyclists through a commented map, it may have made later commercial competitors seem more legitimate, especially Google Maps, which eventually began to collect cycle route mapping in 2017. The Google applications provide cycling routes that do not require much input from the users. So the civic-minded data citizen imagined by CycleStreets, who volunteers data and advocates for transparent processes of building applications, is displaced by the consistently data-producing Citymapper customers, who benefit from better optimized experiences of navigation.

ENTREPRENEURIAL CITIZENSHIP IS OPTIMAL

Frameworks of optimization often imply modalities of consumer citizenship or entrepreneurial citizenship, where citizens are expected to solve their own problems or innovate. However, entrepreneurial citizenship, as we saw in the examples of FixMyStreet and CycleStreets, hinges on the same logic as the commercial intermediaries use: creating or rendering data for decision-making and using this data for similar purposes—that is, to visualize disruptions to a city system, in the case of FixMyStreet, and to optimize travel patterns, in the case of CycleStreets. Because these civic actions are directing energy at a narrow range of potential optimizations, they valorize narrow notions of citizenship, celebrating and including people who have the technical skills to effectively employ digital platforms and excluding those who lack access or skills. Even people with skills and access may find cycle maps provided by Google (which uses algorithms developed by Urban Engines and tested on proprietary GPS data pulled from the phones of people using Google Maps) easier to use than the clunkier CycleStreets map based on crowd sourcing. As users take advantage of the more-refined Google maps, more data moves from a civic platform to a commercial one. Communication scholars Mark Andrejevic and Mark Burdon characterize the "sensor society" as a social formation where "the once relatively exceptional and discrete character of monitoring becomes the rule, and when the monitoring infrastructure allows for passive, distributed, always-on data collection."[33] The infrastructure of passive data collection supports the positioning of the citizen as producer of insights for consumption by the government. The data collected generates value for whoever processes it, while the results begin to define ideal ways to engage with the city: where to drive, which route to travel, which litter to pick up, which potholes to repair. As very large companies like Google/Alphabet acquire more technologies like Urban Engines, they also acquire the capacity to frame the use, intentional or not, of data-based applications as a legitimate form of civic participation.

When we assume that cities should be networked and citizens should claim rights of access, we create the underpinning of a form of techno-systems thinking based on the assumption that data collection is necessary for optimizing systems. Such thinking further creates expectations that corporate intermediaries should take the role of creating and managing those systems. The result can be a positioning of good citizenship as participation in adding input to data-based systems. In this vision of the smart city, cloud providers and data processors not only hold the structural power related to control of information flows but have a significant influence on the definition of what should be considered useful or generative data. The definition comes not only from the breathy marketing discourse telling city governments how much better and cheaper things will be when their intermediating platform has been installed but also from the kinds of platforms being built and the kinds of assumptions made in the course of the data analysis.

Given these kinds of platforms and the assumptions associated with them, given the way the relationship between people, the state, and commercial intermediaries is constructed, what kind of citizenship then becomes possible? As more systems move online, the dynamics of optimization—what it includes and excludes—become more evident. The very concepts of openness, transparency, and participation—key features of civic technology activism—have ideological freight accumulated through their relationship to the technology industry. Platforms are shaped by actors with civic interests, and open data activism intervenes in the flattening and commodification of datafied citizenship.

CHAPTER THREE

Entrepreneurial Data Citizenships, Open Data Movements, and Audit Culture

s platformed logics change how citizens relate to the city and the state, activists and advocates find themselves using the same frames and arguments as the commercial intermediaries whose data brokerage underpins the move to a big data optimized city. The new logics change the ways that actions are assumed to accrue value and advance the value and significance of optimization of personal behavior and collective decision-making. Optimization fits into what some scholars have described as a "calculative rationality."[1] The data-based city and some of the associated data citizenships are bound up with another set of values: participation, transparency, and openness. Data-based platform cities don't only produce cybernetic machines; they are also imagined as places where citizens can gain access to information, hold governments accountable, and use information as an open resource that allows everyone to participate.

The promise of optimization (taken this way) is that data can also help citizens to gather and share relevant knowledge and to participate in urban governance. That is the driving assumption behind advocacy movements that intervene in the platform governance movement not by collecting their own data but by using the principles of data-based governance to advocate for transparency and open access to information. They use their own participation not only to open up data but to call it into question. Open data movements have spread across the world in the past decades. They have driven national policy at the highest level and are bound up with the shift to platformed governance. Listening to open data advocates talk about their work and observing how they have questioned both data and the way that it is used in governance decisions show the potential of these movements but also some of their constraints—including how difficult it is to challenge ideas about platform governance that come from narrow parts of technology culture.

DATA AND THE STATE

Urban governance has been solidified through local government's collection of data.[2] Rob Kitchin and his coauthors identify how city performance indicators make meaning, noting that the indicators employ a limited instrumental rationality and promote a powerful realist epistemology. They write, "For their advocates, the power of indicators, benchmarking and dashboards is that they reveal in detail and very clearly the state of play of cities. They enable one to know the city *as it actually is* through objective, trustworthy, factual data that can be statistically analyzed and visualized to reveal patterns and trends and to assess how it is performing vis-à-vis other places."[3] This instrumental rationality underpins forms of new managerialism that reward adherence to targets and discipline underperformance. Such rationality also opens up the space for intermediation in establishing targets and future goals as part of a process of "anticipatory governance" that depends on platform logics because they require the integration and processing of data to determine possible directions. Governance, rather than being responsive, uses data to create possible trajectories and uses analytics to judge the possibility of how these might play out. Louise Amoore calls the resulting juggling of priorities the "politics of possibility."[4]

As outlined in the previous chapter, the governing assemblages that Kitchin and his coauthors identify create ways of framing and concentrating on citizenship that are oriented around instrumental rationality and embedded in various versions of new public management and neoliberal discourse.[5] With the rise of data analytics systems, automated decision-making systems like machine learning can be applied to large-scale, variegated sets of data produced by citizens who accept that their movements, buying habits, energy use, and biorhythms are collected as implicit participation in the civic project of expanding efficiency. When taken further, this kind of acceptance permits governments to roll out facial recognition systems or other data-based systems that depend on constant monitoring and profiling and deliver biased, dangerous, and unreliable results.

If good citizenship requires acquiescing to data collection and trusting intermediaries, are there other positions? Could citizens draw on the ways that discourses or visions of transparency, openness, and participation have been connected to the functional processes of data collection, analysis, and action? In this chapter I show how these processes work to establish spaces for civic action and how open data advocates step into these spaces. Some of these advocates perform the good citizenships that the corporate discourses of optimization evoke, while others develop ways of calling into question how values of openness, transparency, and optimization are conceived of in dominant discourses. These questions are increasingly important as we begin to better understand how data sources can be biased and how coercion into accepting data-based intermediations can exacerbate injustice.[6] If the best way to push back against this bias is to become involved in collecting data, it is important to understand its possibilities and limits.

OPEN DATA MOVEMENTS

A worldwide movement calling for the opening of government data sets has emerged in the past decade and a half. Organizations like the Open Knowledge Foundation in Germany, the Open Data Initiative in the UK, and the Open Government Initiative in the United States have linked with national governments and with nonprofit organizations, including Code for America, to advocate for providing expanded public access to data collected by public authorities. In the United Kingdom, open data and open government spurred the restructuring of all of the government's digital services. Alongside these top-down efforts, open data movements also included bottom-up participation. In 2007, I attended a typical open data event in London. Sponsored by a technology company, the event was an "unconference" where participants designated discussions of interest and met with representatives from local and national governments to negotiate how to get access to the data that the advocates most wanted to investigate. Discussions in the meeting rooms and corridors hinged on opening up systems and opening up government. So much of this celebration of openness seemed familiar to me—I had heard the same arguments before about open networking systems and their power to allow people to communicate—but this time, instead of the openness being connected with open access to information over the internet, it was connected to an invitation to examine data produced by governments. Entrepreneurs circulated with ideas about how to make citizens understand their cities by gaining access to data. Maybe, it was hoped, digital data could be a material for local knowledge as well as the fuel for optimized decisions.

There is a connection between, on the one hand, open data movements and other open movements advocating for access to knowledge within broader frames of communication rights and, on the other hand, open-source software development.[7] Though concerned with access to information, open-source movements are also focused on an apparent democratization of innovation—a capacity for more people to use openly available information resources in different ways. For this part of the open movement, the significance of open information is not only the availability of different information resources but the ability to innovate and create financial value from it—an assumption that seems to fit with the reinterpretation of communication rights as rights to purchase access or rights to data-optimized municipal services.

Like open-source software production, open data processes suggest a democratization of innovation, with more individuals and small companies drawing on a shared commons of knowledge. Citymapper, mentioned in the previous chapter, is one of many companies that drew in open data from Transport for London; other data sources accessible through the London Data Store became the materials behind artistic projects like the Southwark Tree Map, which uses open data on tree species and locations to produce an evocative map of the trees across that London borough. Open data is also behind engaging and dynamic representations of cycle traffic across the city based on the docking in and out of rental bicycles.[8] The London Data Store is one of thousands of similar stores created in cities across western Europe and North America, taking the form of websites with links to available data. The websites usually allow visitors to click on and download files of data—sometimes using the .csv format that can be processed on a spreadsheet and used to create illustrative artworks, but often in PDF form, permitting reading but not any manipulation of data. Despite the limitations in the design and accessibility of data stores, many local governments have promoted and supported the production of third-party apps using this data, often by hosting events like the one I attended and, later, hackathons that provided kudos and prizes to groups who could build new apps from open data.

SOFTWARE DEVELOPMENT FRAMES

These activities and frameworks draw strongly from the culture and practices of open-source software development, where access to shared digital information in the form of software source code makes it possible for people who know how to program to create other new software resources. Data, advocates thought, might be able to work the same way and create an open-innovation ecology of data reuse. Along the way, a version of openness descended from software production culture connected with a rapidly growing data-based economy and a rapidly shrinking administrative state to produce a new kind of good citizenship in relation to data: citizen auditors who could hold government to account.

The most common definitions for "open data" are functional, emerging from a working group convened by software developer Tim O'Reilly (who also coined the terms "Web 2.0" and "government as platform"). Based on principles set forth by the working group, open data is defined as being complete, timely, accessible, able to be processed by a machine, nondiscriminatory, available without registration, nonproprietary, and free of copyright or patent restrictions.[9] The implicit assumption is that such data are fully accessible, in a usable form, and, like computer code, can be adjusted or made to work in new and different ways. Of course, this work can happen only if the data are made accessible in a particular form and if the people who access it have the required knowledge and skills. The ideas embedded in and associated with this form of openness depend on high levels of expertise, as well as comfort in the high-tech cultures associated with Silicon Valley. In practice, many of the people whose lives are impacted by software-based decision-making processes are excluded, including those affected by software used for sentencing (for example, COMPAS in the United States) or predictive policing.[10]

Regardless, policy interpretations of the potential of open data suggested that it might also enable participation in governance in new ways—to bring practices and sensibilities from hacker culture into advocacy, to open up different spaces of participation, or bring different perspectives and expertise to the use of government data.[11] In the United States, at the beginning of the Obama administration, for example, an "open data directive" committed to principles of "transparency, participation, and collaboration" suggested that citizens should help to make information and decisions at the state level transparent by working with open data and that both states and citizens should restructure their relationships to facilitate participation and collaboration. In the United Kingdom, beginning in 2010, the Conservative government (first in coalition and then in a majority government) advanced a platform that included an increased focus on transparency.[12]

In the United States as well as the United Kingdom, the data citizenship associated with early open data advocacy drew on ideas about the ability and willingness of citizens to audit government data and present it in new ways. Although this auditing is more active than citizenship performed by constantly producing data, it is also shaped by commercial and platform logics. Auditing citizens were initially imagined as holding states to account by enforcing transparency and thereby facilitating the move toward smarter governance.[13] However, advocates within open data movements in the UK interpreted their own role as extending beyond transparency; they wished to create space for critiques of received knowledge that went beyond auditing. Once advocates started to understand the platform environment, they pushed further still, to rebuild and intervene in the construction of civic data platforms. This kind of data critique we might think of as ontological, a philosophical concept that I define as relating to the structure and order of things as they are. In order to unfold how this occurs, we can begin looking at how transparency and auditing connect to the processes of data-based governance and the move toward platformed governance.

OPENNESS AND THE AUDITING CITIZEN

Positioning auditing as something that good citizens do links into the idea that governance depends on being able to see governable entities and render them transparent using forms of knowledge that allow governments to see them through mapping, official statistics, and other knowledge technologies. This is what anarchist historian James C. Scott calls "legibility." Scott writes, "A thoroughly legible society eliminates local monopolies of information and creates a kind of national transparency through the uniformity of codes, identities, statistics, regulations, and measures."[14]

In the United Kingdom, the popularization of the active citizen is connected with the expansion of transparency advocacy during the Conservative–Liberal Democrat coalition government that served from 2010 to 2015. In 2011 the enthusiasm that I observed in the open data meetup a few years earlier had wound its way into government policy discourse, where it began to be used to describe some of the ideal features of citizen-advocates. Eric Pickles, then the communities minister, argued in favor of opening up the data created by government, saying, "Greater openness in spending is the best way to root out waste, spot duplication and increase value for money. That is why I have been asking councils to 'show me the money' so local taxpayers can see where their hard-earned cash is going." One key part of Pickles's plan was for individuals to scrutinize spending information and hold their governments to account. In the press release, he is quoted as saying, "The simple task of putting spending online will open the doors to an army of armchair auditors who will be able to see at a glance exactly where millions of pounds spent last year went. The public and the press can go through the books and hold Ministers to account for how taxpayers' money is being spent."[15]

The phrase "army of armchair auditors" evokes empowered citizens who possess the expertise and motivation to query specific acts and decisions undertaken by the state. It also presumes at some level a lack of trust in the state and a reliance on the expertise of individuals to hold public institutions to account, particularly through scrutiny of their actions. This form of governmentality is individual rather than collective, and reactive rather than constructive. Civic responsibility becomes expressed in terms of "speaking truth to power" by creating alternative or oppositional narratives. Data auditing can speak truth to power, but more significantly, it can create oversight capabilities that once might have been the responsibility of the government itself. Auditing citizenship transforms civic action into data processing without fundamentally disrupting the frameworks that valorize predictability and data production.

Another set of policy discourses has focused on how the open-source practices of sharing and reusing data could transform government and bring governance practices in line with the tech-industry model of the platform. For example, when the Open Data Initiative (ODI, the UK's arm's-length open data advocacy organization) started operating, it drew heavily from open software and open knowledge arguments to draw out its vision for the benefits of sharing government and commercial data. This nonprofit company, limited by guarantee and funded by the government, aims, in its own words, to "connect, equip and inspire people around the world to innovate with data." The ODI's advocacy, which began with calls to render government data transparent, was based on assumptions that the data was produced by the public and so should remain publicly available, especially since it lowered public administration costs. Its second-year report says: "Companies are already approaching us to explore openness as a competitive advantage. Open governments, open cities and regions, open industries and open research are all in progress. Open collaboration and open markets are the only approaches that will scale."[16]

The ODI's foundational description of the culture of openness drew directly on the idea that the peer-to-peer culture modeled by open-source software development could and should be reproduced through activism related to data. The second-year report made that clear: "We have a generation growing up in a peer-to-peer open culture. How will Generation O transform our public, commercial and personal spaces? If collaboration is at its core, should we evolve from user-centric design to culture-centric design? This is a shift of web-scale, a cultural artefact, a shift in our collective psychological awareness. We are all data now."[17]

When we are all data and everything is open, the main points of value and control emerge when data can be collected and managed. While this openness allows advocates to call government data into question, it equally makes it harder to question the foundations of datafication or its consequences. In his description of the benefits of government-as-platform, O'Reilly employs a metaphor of the government as a "vending machine"—set to deliver a service on request—arguing that this is an inefficient mechanism.[18] In comparison to a vending machine, a marketplace platform appears to be far superior, since it allows many different entities to define and contribute to the services being delivered. But seeing the city as only a marketplace reduces citizens to consumers and makes data advocates the critics of an administrative state that technology industry pundits like O'Reilly think should be replaced with more efficient platform-based marketplaces. For example, in the years after the founding of the ODI, it began to advocate for opening up commercial data as well as government data and to act less as an advocate and more as a networked broker of data-related discussions and connections, employing commercial and business justifications to advocate for businesses to release data that could be examined and reused by others. This shift over time illustrates how it is often easier for open data to be understood and justified in relation to innovation and open markets than it is to conceive of it as a shared or collective resource. This difficulty reappears in discussions of whether, when, and how to think of data as a commons-based resource.

OPEN DATA MOVEMENTS WITHIN AND BEYOND AUDITING

Open data advocacy helped to bring forward the ideal of an open, market-based governance platform, in some ways to the dismay of open data advocates themselves, who often had a more nuanced sense of the potential for open data to act as a commons resource. Jo Bates, a data studies researcher, performed a study of the UK open data movements in the early 2000s and identified how open data advocacy came to serve government interests. Drawing from political theorist Antonio Gramsci's theory of *trafismo*, she argues that many of the potentially disruptive and critical positions of open data advocates were repositioned to serve government interests. In particular, she identifies how open data policies unfolded in the UK against a backdrop of consumer citizenship and increasing private provision of public services. The rise of open data also aligned with more technocratic decision-making, a reduction in state involvement in welfare, and a diminished framework for democratic decision-making.[19]

Celebrations of openness and transparency appear in policy discourse and also shape the frames for civic organizations and model the ways data advocates are expected to act as good citizens. Like the community-networking advocates who used visions of the smart city to advance the prospect of community-owned infrastructure, data advocates use the language of openness and platform-based optimization to advocate for the production of public information infrastructures and the transformation or recapture of the function and consequences of platforms.

Even when advocates take the position of auditors examining government data, they do not necessarily assume that raw data is perfectly truthful; rather, they assume that having open data opens up a space for interpretation. The result could be a generative engagement with data: "sharing raw data makes the process of interpreting it transparent and breaks governments' monopoly, which means that everybody can make his or her own interpretation of the data that governments use to make and justify their decisions." Some advocates see this as a democratization of information, in line with the democratization of expertise that political scientist Beth Noveck suggests is the foundation of platform-based smart citizenship.[20]

Interviews with open data advocates suggest a more nuanced understanding of both democratic engagement and the capacities of data-based knowledge. Calling for transparency is not a means of getting at some ground truth but instead can serve as a means of making the process of using data to construct truth and validity transparent. "Sharing data in 'raw' form—'as collected'—is not about revealing an unbiased and objective truth, but about making the biases of this data transparent and allowing 'more interpretation of truth.'"[21] Interviews, observations, and other focus groups with open data advocates illustrate how they employ the themes of openness and transparency, sometimes explicitly using the same language as the government and commercial operators while also raising new opportunities to use data as a material for civic action.

DATA IS NOT TRUTH

In the southwest of England in the region connecting Bath and Bristol, open data advocates started a group called Bath: Hacked in 2015. It began as an informal meetup of interested citizens and is now structured as a community interest company supported by the local government and by the data-hosting platform Socrata. The organization meets regularly for hack days and maintains a portal to access local data on economy, health, education, government, environment, heritage, population, and transport. There are 123 datasets and 110 geographic datasets. The organization's website also hosts numerous data visualizations. Bath: Hacked is known to be one of the best-organized and most active local data organizations in the UK, being especially effective at linking open data to important local issues and mobilizing community participation related to data. By 2016, Bath: Hacked advocates had started to think about their role as being less about collecting and maintaining data and more about interpretation, management, and stewardship of open data resources.

In 2016 a fellow reseacher and I met these open data advocates at a series of hack days and open data events in Bath, and over the following years we encountered them at other open data events across the country.[22] Compared to the community Wi-Fi activists, the open data advocates were a bit older, and many had worked in the public sector. There were more women, but the majority were male. Their way of discussing their concerns was also more embedded in government processes than was the oppositional, hacker-inspired language of advocates for distributed wireless networking. The Bath: Hackers felt a serious responsibility to use data and their skill in interpreting it to draw out different political arguments and presenting them to people. Enthusiasm about their responsibility to bring issues forward was tempered with concern about how their expertise might play out across other contexts. When we asked about the importance of maintaining a public website, one of the main developers replied. "People can misanalyse the data and that's partly why we wanted this website. . . . I think if we graph stuff properly and have the right context and background information associated with the data, then people won't misinterpret it, or they should give themselves hints as to what is legitimate use and what isn't," said this Bath: Hacked member at a group interview in Bath in 2016. Another member, who had worked as a consultant and volunteer with small organizations helping them to use open data, reflected on the balance of knowledge and risks of misinterpretation: "[In] the kind of work I've done just helping organizations, helping with data, I've come against that kind of defensiveness. [I'm] often worried about risks of misinterpretation and use."

As these comments illustrate, some of the most active data advocates regard their capacity to use data to solve problems with a certain amount of ambivalence. Even when data is available and expertise can be found to interpret it, the responsibility for bringing concerns forward sits between the individual and the collective. Simply bringing forward data does not bring forward alternative perspectives. For example, the Centre for Investigative Journalism revealed that local advocates living in the Grenfell Tower in London (which caught fire in June 2017 owing to the installation of illegal and flammable cladding; seventy-one people died) had accessed open data related to their building before the fire and documented some of the results of their investigations in a local news website.[23] In this case, the publication of the open data and its analysis by the local residents had failed to force the government to review standards for building cladding. Despite years of data activism identifying problems with the buildings, the voice of the people was heard only when their warnings were not heeded and their home was consumed by fire; some of the people involved in identifying the relevant data died. Faced with the reality of class-based and racial discrimination, data do not automatically generate the kind of stories that garner attention, and data that questions the decisions of the powerful can still be ignored.

The open data activists in Bath encountered similar issues. In 2016 they worked to identify electricity use in public buildings across Bath, but as specific data were released, the advocates became concerned that publishing it might mislead people about electricity use and local government expenditures. The data alone could not tell a story; it was open to interpretations that might have ideological consequences. One interviewee reflected: "I guess my concern would be, you've got organizations like the Taxpayers Alliance in the UK who are very anti-waste in the sort of public realm and they might look at electricity as an issue [saying,] 'Oh, that's absolutely terrible. It should be zero. It should be switched off.'" But, as he pointed out, someone who didn't know services had to run overnight might "get the local paper to write an article saying that the [local government] absolutely wastes all this energy . . . and then the council will be on the defensive."

Being a civic intermediary for published open data is a significant responsibility, requiring advocates to strike a balance between the risks of publishing and the benefits of transparency. Organizations, including governments, can be defensive about making data publicly available and worried about misinterpretation. An interviewee said, "The way you deal with that is make sure you publish all the context that goes with it. So, you describe how it's been collected. You want to help people understand where there are limits in using the data, ways that you can usefully use it or ways that you shouldn't use it." The assumption that opening data and inviting participation create collective intelligence starts to unravel when advocates talk about the details of their experience. Data can inspire debate, but people have to be willing to analyze it, which must be done carefully.

"If the data isn't there and nobody's analyzing it, then everything's just opinions," explained an interviewee. "So, you kind of want people to be engaged in issues so if you're giving them the data and then they actually start to use it then that feels like a step forward. Because it kind of shifts the debate to talking about facts and figures rather than just opinions." Advocates are very clear that open data are beginning conversations about truth and reality rather than being truthful or representative in and of themselves. Instead of holding the government to account by revealing excesses, advocates open up questions and discussions, which, as the Grenfell Tower tragedy illustrates, may still be dismissed because they come from people whose voices are already marginalized.

Simply providing open data can be problematic, as the Bath: Hacked example shows. Many open data advocates even question the idea that more open data is better. One advocate reported that he wanted to avoid "the really unhelpful outcome you get from most engagements with open data": you're "left with more questions about what the data means than answers to the question you approached it with." Most open data advocates have had to develop their local civic practice against the backdrop of rhetoric celebrating the value of open data for shining light on the excesses of government. As these comments reveal, open data advocates, even when they take on the role of interpreter that platformed government affords, are aware of the politics of information and the ways that their work legitimates or delegitimates the institutions of government. One of the Bath: Hacked advocates reflected wistfully that "governments wanted to get more money out of this or good PR or, I don't know, or a mixture of these things, and civil society was hoping that there was going to be more trickledown, and less corruption. But I feel like probably most people agreed that neither of these happened to the extent that they originally set up or envisaged."

These ways of thinking about open data tie into a longer history of using statistical data as a form of critique. Isabelle Bruno, a historian of statistics, and her colleagues refer to this as "statactivism," which they define as "utilizing diverse methods of quantification so as to produce groups, subjects, that arise from an aspiration to liberate themselves from conditions to which they are beholden." Such projects both represent and critique reality. The notion that statistics have the "authority of facts" is used to "show the possibility of an aggregate reality other than that put forward by the institution."[24] Similarly, being able to access open data creates the means to question validity and institutional power by encompassing new categorizations and modes of reality while also questioning some of the foundational assumptions of the platformed big data city.

INTERVENING IN THE DATA PLATFORM: STANDARDS ACTIVISM

Between the optimizing systems and the data volunteered by citizens sit the protocols for managing information flows. These include data commissioning standards. A global community of data advocates has started to focus on the standards that control how data becomes open. One of these advocates is Michael Lenczner, the community wireless advocate from Montreal, who now works on setting global standards for access to data about philanthropic giving. Another is Tim Davies, an advocate of open government data who has built a small business that focuses on setting standards to ensure that government services purchased from private companies provide open and shared data. A third is Greg Bloom, who dedicated many years to developing interoperable standards for sharing community service data. Their projects bring forward a new critical perspective on open data, one that generates new organizations and structures for open data as an alternative ontology. Their projects also highlight how these types of activism depend on expertise and social position.

As governments move toward opening platform-based data to civic scrutiny and competition for contracts, their goal is not so much to make themselves accountable as to make them efficient. How can advocates intervene in this process to make these processes accountable or just? Figuring this out was one of the motivations for Tim Davies, who founded the Open Data Services Co-operative after studying open data movements. He thought a consultancy would help bridge the gap that he identified between publishing open data and creating the expected social impacts. In particular, Davies's project has tried to maintain the notion of openly accessible information in a context where public services are delivered by public-private partnerships; he has come up with standards to ensure that data used within or generated by these partnerships might be publicly accessible. Although his work started in the United Kingdom and Lenczner's in Canada, both projects have taken on global reach.

Specifically, Open Digital Services Co-operative works to establish the standards that governments use to open their data. The standards have the potential to specify how data is used, how it is structured, and what rules are applied in using and reusing it. Unfortunately, enthusiastic support for the idea of open data among governments has led to an increase of work and influence for a small number of intermediaries. As Davies pointed out to me, even though rhetoric stresses the democratic potential of open data, "there are a relatively small number of intermediaries and a relatively small number of data sets ever get mediated in that way. There is still a broad theory of change that says, 'We'll put out the data in more formats. It doesn't matter if it's not terribly accessible because other people will do things with it and then we'll get use at the end.'"²⁵ This perspective imagines open data to be a marketplace where individuals can compete with each other for the possibility to build a new product from it.

To address the limits of this narrow view of accessibility, Davies and his colleagues focused on doing the technical work of organizing and reviewing standards as a way of ensuring public access to information and improving transparency. They understood technical standards as a kind of "Trojan Horse" for policy advocacy, using standards to advance their goals of increasing accessibility, reuse, and the use of data commons. As Davies explained, "The standard is acting as a way of encouraging people to implement . . . reform. If you'd gone to them and said, you must implement this reform to keep an audit trail of all your contracts because we are worried about corruption, you'd have got much more political push-back than you seem to get through saying [you should] adopt this standard."²⁶ The contracting standards that Open Digital Services Co-operative created include standards intended to prevent data from suddenly becoming the property of a different company after the takeover of a government service-provision contract. The detail-oriented excavation of the issues underlying the brokerage of data in smart cities shows the high level of political and technical skill required to carry public interest requirements into the heart of the function of a platformed government. For community advocacy using data to be successful, that data needs to be accessible, comparable, and usable, and it must remain in the public realm. For data advocates to speak and be heard requires that they have significant comfort with techno-systems thinking.

The standards work of Open Digital Services Co-operative offers ways of bringing back particular forms of data into public ownership or consideration and of bringing civic advocacy perspectives into policy-making conversations. By intervening in the structures that provide access to data, standards work provides material for decision-making. As Davies comments, "A number of interesting things happen when the data is more available. It is easier then to work with outsiders to build tools, build things, discuss processes, get ideas . . . we find ourselves through the open contracting help desk on calls with government where we are discussing wider reforms to their systems, in ways that if we'd just said we are an advocacy group who want to promote more transparent procurement would not have happened."27

OPEN REFERRAL: STANDARDIZING DATA, ADVOCATING FOR DATA COMMONS

Through my involvement with the UK-based open data community, I met one open data advocate who wanted to go even further than commissioning standards and try to reorganize civic data using the model of a commons. A man much more in the hacker mode than Tim Davies, Greg Bloom had worked in community food banks and on grassroots civic projects in Washington, DC, and San Francisco. Between 2014 and 2017 he dedicated himself and much of his personal savings to developing a bottom-up community-oriented data intermediary that would collect and distribute data about support services for vulnerable people. On several visits to London, Bloom told me about his project and asked for help trying to imagine how it might apply in other contexts. Bloom started his career doing technology-oriented work in community organizations. He started The Open Referral project when he noticed that within a single city, different community service organizations maintained separate lists of information about where people in need could access services. This made for some inconsistencies: people seeking help from one charity might be referred to a different charity than if they had started in a different place. Even where such lists had been computerized, they were rarely integrated, nor were they released as open data. Bloom imagined a platform for open referral data that would integrate the quasi-standardized information held by community organizations, using web standards to make indexable online. He recalled that the organizations had "a directory database and access. Their old mode was duplicating it and sharing it with all these other organizations manually, the new mode is they want to move into an open database with an open API [application program interface] that other systems can consume from automatically."[28] This new mode turns a set of separate resources into a single resource that many different organizations can access.

Bloom thought that this kind of standardized system could help prevent information from being duplicated, hoarded, or buried. When he piloted the project in San Francisco, he recalls, "within a four-block radius you had the San Francisco Bar Association that produced a 400-page resource directory in Microsoft Word. You had the San Francisco Homeless Wiki, which was run out of this guy's grassroots church, which was basically just transcribing what was in a printed booklet onto a Wiki. Then you had a new sort of Silicon Valley–style agile, lightweight app that also took the Word document and converted it into a database and put a

nice mobile interface on it." Bloom wanted to use the platform-based data-sharing model to make civic information
efficient and transparent: "to get to the point where there are
three organizations and previously they were all operating es-
sentially in silos, now their systems can speak to each [oth-
er]. . . . Like, this one is going to monitor services for elderly
people, this one is going to monitor services for youth, and
this one is going to monitor mental health services and they
have the means to check on how that's going."[29] In these comments it is possible to see the same expectations for monitoring, data flow, and optimization of systems that thread through
Davies's comments on standardization and echo the promises
of the data-based optimized smart city.

Bloom has seen tech startups trying to occupy the space
that nonprofits previously worked within. For-profit referral
services are growing, and many small social enterprises are tar-
geting service data, especially health data, hoping to build
enough value in the development of services that they can be
acquired by information providers like Google. Such systems
provide a useful service, but the data behind them is proprietary and cannot be reused; over time it builds up, eventually
constituting a significant resource outside public reach. As
Bloom writes in a chapter describing his earliest version of
Open Referral, "If these startups evolve into yet another class
of intermediaries, institutionally committed to protect their
hold on data about our communities, the real problem—that
communities lack effective means to produce and share their
own information—may only be entrenched."[30]

Open Referral can therefore be viewed as a type of data
commons. The data commons, like the wireless commons, can
be a way of figuring technical resources as shared sources of

value and significance. But as open data advocates have pointed out, simply making data available and standardizing it is not enough to successfully create a commons. Community resource databases, for example, are freely available on GitHub, but without agreements and standards for using the data within those databases, they do not function well as commons. Drawing on the work of Peter Levine on information commons, Bloom suggests an "associational commons." It consists, he says, "not just of the resource itself (the open field of grass, or the open set of data) but also the synthesis of that resource with the web of social relationships that form around it."³¹ The relationships are structured in part by the kinds of standards and frameworks that Open Digital Services Cooperative pursues.

Bloom's conclusion on the necessity of a public data commons, where rights to access and use information are negotiated by rules and standards, may be a productive critique on the inequities of the platformed big data city. Much depends, however, on how that commons is defined and whether it focuses on an individual or a collective data subject, employs a platform model that opens out significant roles for intermediaries, or seeks to reposition and legitimate different forms of knowledge, relationships, or care within urban space.

TRANSPARENCY OBSCURES INJUSTICE

The platform model for governing the big data optimized city employs values of transparency and openness to legitimate an information politics that calls state legitimacy into question and that leaves open spaces for civic knowledge. The caveat is that the idea of the city as a platform frames how this supposedly transparent open data should be used, including by entrepreneurial, auditing citizens who exercise rights claims by holding the government to account. The caveat raises serious questions about when and why data-based activism fails. In the case of the Grenfell Tower fire, people with the responsibility to mitigate risks ignored the data collected and published by residents. It is likely that the social position of the residents as social housing tenants influenced how their data were interpreted and whether their voices heard. If data activism requires, at a minimum, expert knowledge as well as a social position that gives volume and immediacy to the advocate's voice, critiques will be limited. The difficulty is exacerbated by the way that many data activists are absorbed by the apparatus of government. Even the most experienced open data advocates acknowledge that the promises of openness that inspired participation by expert amateurs have created new forms of power. Collaboration and participation on city platforms have produced a cadre of linked-in experts and consultants who can move between advocacy and city administration.

In my interactions with data activists, I was most impressed by the activists who sought to employ their specific expertise to transform the systems through which data is used and controlled. Forging standardization agreements that allow data to be placed in commons and creating agreements that specify transparency and public ownership of data even in big data optimized cities are unique and possibly transformative interventions. The civic technology movement, which technology researcher Andrew Schrock has investigated, draws strongly on the idea of a "civic hacker." Civic hackers can, for example, be invited to a hackathon to use government open data to solve problems defined by the government. Hackathons have been critiqued as exercises of entrepreneurial citizenships—exercises that produce reactive and uncritical responses.[32] This attitude is part of the ambivalence associated with activism performed by well-resourced, confident, and expert participants. Paradoxically, the same expertise that allows open data advocates to intervene in the construction of platforms also draws them into positions of power in the same institutions they critique. Tim Davies reflects that "relationships between civil society technologists and government become very friendly," making it harder for the activists to make strong claims. "It may be now that the door is open to collaboration you don't need to do as much claim making, but I think it also has risks of corruption."[33] Understood in relation to communication rights, the ambivalences inherent in data activism limit who can, and how they can, transform a narrow data-based auditing of the state into something more transformative. The principles of openness and transparency bring their own additional risks.

As the frameworks of neoliberal governance give way to something darker and stronger, and as sanctuary citizenship is undermined by the control of human beings as they attempt to cross borders, the value of openness seems to have been broadsided by what appears to be the consequence of openness: the development of many systems of measurement that hinge on practices of sharing and participation and that produce dataveillance as a normal state of being. Open data advocates, with their careful reflections on data, truth, fact, and relative power, do not intend to intensify dataveillance. Underpinning their interest in transparency and collaborative participation in making sense of data is an idea that data needs to be available and that increasing the amount of data measured makes it possible to make accountability central to public functions. However, both enhanced data collection and enhanced openness support the development of publicly accessible information infrastructures. What should these infrastructures look like? Greg Bloom of Open Referral calls them commons, and so do others, but they don't always mean the same thing.

CHAPTER FOUR

Rethinking Civic Voice in Post-Neoliberal Cities

I N today's smart-city visions, we hear how small connected sensors like air quality monitors and traffic cameras can be connected and programmed to generate data that optimize city life. For citizens, the promise is that this accumulation and processing of data may allow civic action to be directed in new ways—bringing forward data about things that matter and enhancing the voice of citizens—beyond simply analyzing government data. The idea of voice comes from media theorist Nick Couldry's work about the possibility for people to speak or be heard on issues that matter to them. In our joint work, Couldry and I wondered how sensing or tracking data might be mobilized in service of this collective voice.[1]

Collecting sensor data, including environmental data and data on air pollution, appears to provide a newly effective way for citizens to draw attention to local issues. Homemade open-source sensors were used in the community-based mapping of the Deepwater Horizon oil spill in the Gulf of Mexico in 2010. After the spill, the citizen science organization Public Lab taught people how to build sturdy balloons mounted with digital cameras that could float above the spill and record its impact from a different perspective than the company's when it took measurements. Citizens' photographs of the spill circulated, influencing public discussion, with some commentators arguing that the citizen narratives led to increased accountability from the company, BP.[2] This was one of the first projects to foreground the capacity of citizens to collect the data that they chose, rather than having data passively collected and "crowdharvested" from commercial organizations or governments.

A generation of civic sensing projects have been undertaken under the assumption that creating and interpreting data can allow citizens to produce different—oppositional—knowledge from that produced by authorities and experts. However, sensing as a civic practice still continues to support the platform model of the city. This happens even though sensing can be a way of bringing out new knowledge and supporting multiple ways of knowing—something that sociologists Engin Isin and Evelyn Ruppert consider to be important aspects of digital citizenship. This creates structural constraints on how data is able to directly create voice. It may be the case that a complaint or concern becomes perceived as more objective when it is presented using digital data. For example, consider the discussion by Mara Balestrini and her coauthors of noise sensing in Barcelona. Residents living near an area popular for late-night drinking complained of excessive noise, but their protest was dismissed as "individual perception." Coordinated use of calibrated noise sensors allowed the residents to argue that the noise reached levels considered unacceptable according to public policy.[3]

These civic efforts to use data unfold within a particular context: often a city where fragmentation and privatization of infrastructure have been intensified by efforts to acquire, manage, and employ as many forms of data as possible. Sensor data can be used to trigger action—for example, a full recycling bin trips a sensor that records its location so it will be included in a future drive-by. For sensors installed by governments or companies, the relationship between the sensing data and the action remains obscure, raising questions about whether the sensing or the data processing is the more significant site of power. Citizen action using sensors can often mean that citizens are sensing things that governments or corporations won't. If decision-makers view it as legitimate, this data from the bottom up can be used to replace city services by way of temporary contracts with digital intermediaries. When the data is not viewed as legitimate, it can be dismissed out of hand.

Digital inequalities emerge as a consequence of the intensification of information control by intermediaries. Stories from a pilot project in Bristol, England, suggest new possibilities for data commons to challenge this control. In Bristol, data commons, rather than being only stocks or stores of citizen-generated data, became sites to develop relationships of solidarity through the tolerance of friction, tension, and dispute about how data connects with things that matter to people. Control of information and its predictive power creates new kinds of inequalities, but data commons also have the potential to become spaces for discussion and struggle in relation to governance; looking at them here will illuminate the challenges of addressing civic problems when decades of privatization of public services and the push to a platform economy have captured not only data but the possible spaces for predictive decision-making.

DATA-BASED SPLINTERING OF SERVICES AND KNOWLEDGE

In this book I have so far described the commercial interest in creating multisided data markets based on pervasive personal data collection and data analytic processes so that city governments can purchase insights or predictions that help reduce risk. These tendencies did not emerge in a vacuum: they built upon processes that have unfolded over decades to shift public urban infrastructures into private operation, a process that geographers Stephen Graham and Simon Marvin describe as "splintering urbanism."[4] Splintering processes break up the common infrastructures that serve cities, such as water, electricity, and internet, and allow different operators to deliver them in different areas. This leads to a differentiation of their quality: a "redlining" of services in poorer areas or in line with racial bias. In the case of data-based optimization, splintering intensifies, but instead of control of a resource, "datalining" separates out service qualities based on the amount and richness of data available and the capacity to manage, broker, or sell that data. Datalining is embedded in the overall transformation of cities from being providers of public infrastructure to being brokers of service contracts from many private operators. The process of separating out control of infrastructure has been replaced by a process of managing flows of data and the intermediaries who define and control it.

Arrangements are required to put this vision of a city into practice, to transfer expertise and oversight away from municipal governments and toward the global corporate actors who retain data-processing capacity at a large scale. Google, for example, has mapped cities and, by acquiring real-time traffic-sensing apps like Waze and Urban Engines, now holds the capacity to predict traffic based on real-time sensing of the locations of pedestrians, cyclists, buses, and drivers. Here, managing data flows provides another, different layer of splintering and control. Similarly, route-planning applications and smart electricity-metering systems depend on real-time data. In the pre-splintered city, the local government might have owned or directly controlled the bus company or, as in Fredericton (discussed in Chapter 1), might own and control internet infrastructure.

Except in increasingly rare examples, city governments now contract services from third-party companies for bus driving, water provision, and database management and have little control over how they are delivered except through service-level contracts (which is one reason why the Open Digital Services Co-operative seeks to build protections for public data into these contracts). Third-party contracting for data collection, management, and analytics takes an enormous amount of information about citizens away from the government. Not only city services but also the experiences of citizens—as well as the potential to interpret their future needs—begin to move out of public responsibility. This move is not inevitable: in some cities, including Seoul and (until recently) London, data collected from RFID-enabled transit cards was considered public data and was used for public transport planning based on analysis of demand.[5] However, as data-based intermediaries like Uber capture important parts of the transportation data space and do not share any data with local governments, data gaps create new splinters. When monitoring data for water use, electricity, waste collection, environmental health, and transport are all collected by different entities who each try to optimize their service provision, there is no longer a way to bring this information together for public benefit.

CITIZEN AS SENSOR: CIVIC VOICE

In the struggle to understand, control, and speculate upon data, citizens have often been positioned as passive "citizen sensors" who generate data for authorities to use. Yet citizen projects have also used sensor-based data collection to challenge the biases that build up through data-based splintering. In fact, citizen-sensed data often appears to be more accurate than the splintered data, since "the observations presumably reflect the lived experience of an individual who 'knows' the place well . . . [participatory sensing] can take on the affordance of truth." Perhaps because of this promise of truthfulness, some sensing projects have focused on making sensing accessible to individuals without really considering how the data generated might be used. For example, in 2012 a coalition of designers and early Internet of Things (IOT) advocates distributed a DIY hardware-based Air Quality Egg and invited egg purchasers to upload their data to an online platform. The Citizen Sense project critiqued the egg, asking, "How one is meant to respond to the glowing egg has been left to the user to decide. The egg can be seen as an open design that enhances participation, although it would be interesting to research when deployed whether it achieves these material and political effects."⁶

Critical theorist Jennifer Gabrys has argued that the capacities of sensing render citizenship into a Foucauldian project where citizenship ceases to connected with the exercise of rights and responsibilities and becomes associated with how well citizens can present computational information. Gabrys writes, "Participation involves computational responsiveness and is coextensive with actions of monitoring and managing one's relations to environments, rather than advancing democratic engagement through dialogue and debate."[7]

In a participatory sensing project at a series of hydraulic fracking sites in Pennsylvania, Gabrys and her team found that the civic sensors weren't as well calibrated as the official sensors, but that by measuring air quality, noise, and water quality in different kinds of locations than the fracking companies or the Environmental Protection Agency did, people generated "just good enough data" to start conversations with policymakers. The project brought air quality data from the team's pollution sensors together with the "eyes on the ground" observational data. Gabrys's team argued that the civic action that was most relevant to their participants came from "making data" that addressed matters of concern and opened up conversations. But what are the risks of the "just good enough"? One outcome of citizen science and civic data collection can be that government agencies like the US Environmental Protection Agency start installing more of their own sensing equipment and use their regulatory powers (at least the ones they had prior to 2018) to put pressure on companies, who often respond to such pressure by collecting their own, differently calibrated sensing data. In Pennsylvania, data-based splintering affected the quality and safety of air and water. Civic data collection didn't so much introduce a new voice as trigger a new data-collection feedback loop.

Almost all of the "greenfield" smart-city projects meant to create neighborhoods from the ground up—by rolling out new systems and building infrastructure from scratch—employ intensive sensing equipment. Some of them fold data-based splintering into their very design, as in the 2016 proposals by Alphabet for its Sidewalk Labs smart-city project in Toronto. The proposals, published in the *Globe and Mail* newspaper in 2019, indicate that the company was testing out the acceptability of operating a private road system, charter schools, and autonomous vehicle systems, as well as individual data identifiers for people and things, in order to "collect a 'historical record of where things have been' and 'about where they are going.'" The company also proposed to limit access to some services when people refused to share data. Alphabet's plans have since been withdrawn, with some commentators speculating that they used the project plans to test public and regulatory appetite for these intensive versions of smart-city data collection.[8]

This version of the smart city places new constraints on civic action. The sensor-based city splinters not only services but knowledge. Each intermediary holds and analyzes its own data. The vision of complete collection and control of services shuts out citizens, and even when they are invited to collect their own data, their data becomes one stream among many. Sensors don't tell stories; by design, they generate well-calibrated streams of data that others can use. If the calibration is off, not only can the validity of the data start to be questioned, but including the data might be impossible in the tightly controlled end-to-end commercial smart-city intermediary systems. What, then, is the role of civically collected data, how good does it have to be, and who gets to use it?

Under the condition of data-based splintering, citizen-collected and commercially collected sensor data become material for training urban decision-making systems, which can contribute to extending and legitimizing surveillance capitalism.[9] Even when the data can be accepted as good enough for the kind of story that can be legitimate in policy discourse, this does not happen without friction. The friction, in fact, may turn out to be one of the most disruptive features of civic or bottom-up data. The data may not fit. It may not tell quite the right story. The friction may emerge in even more obvious ways when the data are collected together or when the resulting data commons are used by different groups of people.

This perspective has something in common with the idea of friction as developed by anthropologist Anna Tsing. Tsing has identified "a central feature of all social mobilizing" that is "based on negotiating more or less recognized differences in the goals, objectives, and strategies of the cause."[10] She argues that misunderstandings within long-term social movements, far from producing conflict, permit people to work together. Tsing uses friction to understand how heterogeneous, un-equal encounters produce energy. She questions the inevita-bility of seamless global flows and instead describes how tense encounters produce relationships of negotiation. These same tense relationships unfold in relation to the struggle over whether and how sensing data can mean something within a longer story of civic engagement and how frictions unfold around its use.

In the center of Bristol, the British city where shipbuilding and maritime trade (including the slave trade) built wealth in the nineteenth century, converted warehouses line the riverside. Some have been repainted and refitted—into the Arnolfini art gallery and the Watershed media center, which hosts a cinema, media production spaces, and the University of the West of England's media production and research center. The last is arranged over an upper floor of the warehouse and includes open-plan offices in the style of a technology startup. Behind the Watershed and up the hill, more water glitters in front of the City Hall, an imposing building renovated in start-up style with modular couches and bare walls that, when I visited, was partly rented out for a private function. Like many cities, Bristol has had to dramatically reduce its municipal budget over the past decade as the result of national government austerity policies, and the rentals and internal reorganization generate cash and reduce costs. It is also, under the guidance of a team of technologists, embracing its own version of government-as platform.

A few miles away, thirty minutes by bus or fifteen minutes by car, the neighborhood of Knowle West unfolds, a mixture of red-brick 1930s council houses and boxy 1960s houses on small cul-de-sacs. Built to house poor Bristolians after slum clearances in the early twentieth century and after the Bristol blitz in the Second World War, Knowle West can be a hard place to find work and can feel far from the center of the city. Parts of Knowle West are among the most deprived areas of the UK, yet the community is also known for being cohesive and dedicated to positive transformation. For the past twenty years, Knowle West has also been known for the Knowle West Media Centre (KWMC), which has addressed social justice issues by providing media training, technology education, and opportunities for creative expression and civic action, all connected through knowledge sharing and technology skills development.

Beginning in 2016, KWMC began working on the Bristol Approach, a framework for integrating socially engaged arts practice into community-based technology projects and smart-city projects, with support from the city council and in partnership with Ideas for Change, a Barcelona-based innovation think tank. The makers of the framework worked with their own idea of citizen sensing as "a process where people build, use, or act as sensors—for example, identifying and gathering information (or 'data') that will help them to tackle an issue that's important to them." Citizen sensing, they said, "is about empowering and enabling people to use technology for social good."[11]

Between 2016 and 2017, KWMC and Ideas for Change tested the Bristol Approach, exploring the potential of a data commons as a tool for social change. In the pilot they experimented with how thinking about shared "matters of concern" might establish a kind of framework for commons-based thinking, long before any data was collected or stored. The project organizers focused on the challenge of collaboratively defining what data should be used for, and tried to temper narrow perspectives associating data with optimized decision-making processes. After the end of the pilot project, many of the same ideas were tested out in the Making Sense project in Barcelona and in the REPLICATE Project, which plays with citizen-led decision-making and new technology in Bristol;

San Sebastian, Spain; and Florence, Italy.[12] I interviewed many of the people involved in piloting the Bristol Approach, as well as other people working in creative technology in Bristol, to see how the project navigated the interest in government-as-platform in City Hall, the austerity agenda, the strong sense of community, and the inevitable conflicts that the project raised about data and its benefit.

The Bristol Approach drew both on KWMC's long experience and on a cultural narrative that celebrated Bristol as a "rebel city" exploring radical ideas and open to experimentation with technology for civic benefit. The project connected KWMC, the technology specialists in the municipal government, and certain arts organizations—although some people I interviewed said that not all parts of the creative technology sector were represented. These practices and relationships strengthened connections that were described to me in terms of Bristol's creative and oppositional culture. My interviewees proudly talked about how much they worked together and how they saw the city as embodying a "rebel" spirit.

Katherine Rooney, who was the City of Bristol's contact point for the Bristol Approach, had been advocating for open data within Bristol for years, leading hackathons and working alongside the Bristol branch of the Open Data Initiative. She saw the Bristol Approach as a way to involve the local government not just in producing data for citizens to use but in responding to citizen needs as presented by new technologies. She had already worked with KWMC and its director, Carolyn Hassan, and saw the relationship between the organizations as a key reason for the success of this approach to innovation.

"We're really, really well networked in the City," she said as we drank tea in the newly rebuilt tech-industry-inspired

City Hall canteen, "and my view, because I have a kind of personal mission to diversity open data and the smart city, is to work through those groups rather than without those groups." Like others I spoke to in the arts and research organizations located in Bristol, she had a very clear sense of the purpose and profile of the city's creative identity. "Our kind of digital innovation, smart cities work, was founded as a kind of partnership with, for example, the Watershed. . . . Some of that smart city stuff will be good, but the Bristol angle has always been to bring in the kind of creative, artistic, disruptive angle to it as well from the outset."[13]

In our discussions of how the Bristol Approach sensor project emerged, Carolyn Hassan, the director of KWMC, told me that an interest in sensing was related to thinking about how it might represent different kinds of knowledge: "[We did a] project called The University of Local Knowledge, where we collected around 950 videos of local knowledge with an artist called Suzanne Lacy [in order] to think about how to challenge hierarchies of knowledge. . . . We also were doing a lot of environmental projects. . . . We were interested in who is the expert, who is the expert around understanding our communities." Hassan has had a long interest in media as a way of allowing people to speak. Trained as a photographer, she spent twenty-five years in a West Knowle building, the Media Centre, first as a small photography studio in which she trained youth in photo documentation and later as a center focused on providing training, research, and creative work with a range of new technologies. She described how she anticipated that collecting sensing data could create opportunities for people who don't usually participate in technology development to become engaged, because sensor data might allow people to address important problems: "not the earlier adopters of technology or the tech enthusiasts," she said. "I'm talking about the people who are experiencing most of their problems that we have in our society. So, I'm talking about people who live in communities that are not thriving."[14]

Rooney's and Hassan's perspectives echo the dominant ideas of civic sensing as a form of participatory citizenship. However, the sensor project in Bristol also included the idea that data collected from sensors should be gathered together in a commons that would generate value for all of those who contributed to it. This idea connects with a broader set of conversations about collections of data as sites of collaborative value and the ways that collections, too, create particular types of citizens. The idea of a commons resonates far beyond economist Elinor Ostrom's description of places like grazing fields and fishing lakes as common-pool resources that generate value for the "commoners" such that they should be managed collaboratively rather than reduced through competition.[15] Communication commons like public television airwaves and the wireless connections to the internet explored in previous smart cities have also been thought of as being at risk from a tragedy of the commons.

To test out a different approach, KWMC members tried to define what kind of common problem sensing might address, beginning by spending time in places where people in the neighborhood gathered. As Mara Balestrini, the consultant from Ideas for Change, remembers, "They did takeaway shops, hairdressers, and then kind of more general conversations. They reached out to people who are already having conversations. . . . It's incredibly labor intensive so it's kind of 'expensive.' It's hard to do well. You have to really mean it."[16]

Some of the initial ideas produced through these in-person consultations included a proposal for noise sensors to map birdsong and reflect on the changing bird species locally and to use sensors to connect to mobile phones and tell the stories of trees or local buildings. The project leaders finally settled on the idea of using humidity sensors to help the Knowle West community, many of whom live in poorly insulated houses, to identify and respond to damp in their homes. According to Katherine Rooney, "One of the main reasons why the damp problem was the kind of one that had the biggest snowball around it was because the rental market is as it is, [so] landlords are getting away with people being kind of willing to live in really sub-standard conditions." KWMC wanted to try to disrupt "that kind of landlord-tenant power imbalance."[17]

Once the "matter of concern" was identified, the project started to focus on inviting technically savvy participants who wanted to experiment with building and calibrating damp sensors at hackathons. After the first hackathon, tensions emerged between people who wanted to work with Arduino sensor hardware and those who wanted to work with Raspberry Pi hardware. Workshops continued, involving Knowle West residents and people from other parts of Bristol, including a designer who created a charming frog-shaped box to surround a humidity sensor that, in its first iteration, would collect humidity data on a small SD memory card located inside the Frogbox. Eventually, the project partners hoped, the sensors could be networked and automatically send sensor data to a shared database. Twelve households volunteered to host the sensors for a period of several weeks, and more hackathons followed at which participants worked with spreadsheets of the data collected from the frog sensors and open data from the local government—including demographic data covering the areas where the pilot sensors had been installed and data about previously reported damp issues. At the hackathons, students, artists, activists, and open data supporters worked together to visualize data. The whole event culminated in a showcase and a party.

COMMONS VISIONS

The Bristol Approach began with the collaborative generation of a shared matter of concern, but in the second phase the commons extended to sensor-collected data and government open data. In an early definition of "data commons," human-computer-interaction researchers Dana Cuff and colleagues described them as "repositories generated through decentralized collection, shared freely, and amenable to distributed sense-making" and noted that these repositories "have been proposed as transforming the pursuit of science but also advocacy, art, play, and politics."[18] Their work suggests that building shared collections of data or information might provide new modes of expression. This is often how sensing projects are presented: as ways of bringing into being new publics who can find new sources of expression by putting data together in commons. As Matthew Tenney and his colleague Renée Sieber note, "The public is considered to be an omnipresent crowd and participation is the digital contribution that enables change in social, environmental, and political environments. It is also becoming clear that the level of influence exerted by data-driven participation will increasingly be evaluated in terms of how big a scale it achieves."[19] Along with scale, civic data commons are perceived as gaining value when the data they contain is amenable to calculation by the entities that manage smart urban systems. The calculation, because of the costs and expertise involved, often falls not to the creators of the commons but to the intermediaries who have this capacity, meaning that the civic voice doesn't necessarily come from communities themselves but instead through the creation of "calculated publics" where interests and concerns are represented through algorithmic processing of real-time data.[20]

If the goal is only to generate calculated publics, then other models begin to appeal, like data "collaboratories" containing not only citizen data but commercial data too. Examples are the proposals from the US think tank GovLab that advocate for the extension of data commons to include "personal information actively shared by an individual or a group; information with potentially identifiable information collected prior to any use; information free from any personally identifiable elements actively shared by an individual or group; and information with no personally identifiable elements shared prior to its use."[21] Collaboratories are spaces for access to aggregate data and use APIs, prizes, research partnerships, and intelligence projects to extend data sharing into commercial spaces. Proposals to create them focus attention on the commons as a place for data to be gathered and shared.

The Bristol Approach attempted to move beyond calculating publics but was still confronted with the practical challenges of how to share data and define or derive its benefit. Balestrini described the way the project defined the commons: "by having an agreed sharing protocol and a governance protocol" to define how to access and combine the data.[22] The protocols or agreements thus enact the commons and mobilize resources for discussion, self-representation, and action. The idea here is to move back to the notion, closer to Ostrom's view, that some of the value of a commonly held resource is developed through the collaborative relationships required to sustain the resource.

The Bristol Approach pilot focused on creating relationships and practices that could put sensing and sense data at the heart of discussions about major issues impacting both Knowle West and Bristol as a whole. In practice, the process and the creation of data commons in the Bristol Approach pilot raised tensions. Some came out in my interviews as grumbles and gripes that only particular arts, community, and research organizations within the city participated in the project and that only the usual suspects had their voices heard. My contacts within the city government stressed that they were interested in moving away from the "fairly trad" approach of hackathons to something broader and more interesting, and that taking a bottom-up approach depended on the capacity of partners like KWMC to motivate people: "The big Silicon Valleys keep telling us that engagement will self-emerge and people will self-organize around these technologies. That is not necessarily true. Engagement requires the work of people like Carolyn [Hassan]. The work of people who orchestrate the engagement."²³

WHAT CAN SENSING SAY? CIVIC KNOWLEDGE IN THE POST-NEOLIBERAL CITY

The Bristol Approach pilot offered a way for sensing to help a collective conversation to surface and be interpreted by an entire community. However, it also revealed points of tension, fracture, and glitch that are both continuous with other technological efforts at enhancing citizenship and specific to big data optimized cities and their splintered informational resources.

Although sensing promises to create a better conversation with some "good-enough" data, the shifts in organization and governance that come along with transforming a city into the image of an agile software platform have social consequences.

In the UK, local governments are responsible for ensuring that residents have safe homes. In the past, if someone complained of damp, a city-employed damp inspector would visit to see whether what was reported was damp, mold, or condensation and either request repairs (for city-owned housing) or force a landlord to address the problem. The damp inspector had the power to define a "serious case" of mold and to enforce action. Of course, if the damp inspector didn't record an issue as a serious case, there was not much anyone could do. After local budget cuts eliminated damp inspectors in Bristol, people in places like Knowle West received instructional emails describing how they could spot damp and identify mold and encouraging them to call a centralized number if their self-assessment suggested they had a mold problem. The frog-shaped damp sensor data speaks on behalf of citizens who might have damp problems, but it doesn't speak directly. Instead it gets entered into a feedback loop of sensors triggering instructions to look at damp, perhaps leading to more use of sensors—without ever determining where and how the condition becomes serious—which challenges the idea that civic data enables a robust civic voice. The feedback loop of sensor data needs to involve not only sensor readings but other baseline data. Instead of one expert judg-ing the severity of multiple cases, particular readings might prompt the council to send information to a particular house-hold, who would then need to prove to the council or to their landlord (using more specific data) the severity of their case.

Sensor data alone is not evidence—and evidence cannot be verified in the same way without a damp inspector. Each individual needs to collect further data about what activities they have been doing at what time to compare with the read-ings on the sensor. All of this data collection depends on the sensors working smoothly and providing readings that are consistent enough to be meaningful—either by triggering re-sponses from the local government or by providing enough evidence that a tenant can prove to a landlord that structural problems are causing damp. In the Bristol Approach pilot, sensor data were combined with other data, including open government data on housing types and data generated and processed at the project's Data Jams, supported by the city government. However, the mechanisms to trigger specific ser-vice deliveries were never clearly connected with these data.

Some research has suggested the possibility that citizen-produced data could be applied to important issues like racial-ized policing, unequal treatment in public services, or unhealthy air and water—but the Bristol example suggests that producing the data is not enough. For data to be used to address issues of justice, processes need to be put in place to make the measure-ments actionable and, furthermore, to link the collection of data to the problem defined in the first place. This is a more complex process than simply collecting data or even working through it in a hackathon; it depends on the community advocacy and capacity building that were part of the effort at building the commons but that otherwise don't fit the city-as-platform framework. As Katherine Rooney reflected: "I think you have to be careful that people don't just say, 'That frogs idea is cool, I want some frogs, and I'll build some frogs, and then I'll do the frog thing [to address a problem].' Maybe that's fine but actually they haven't started with the conversation. . . . How you do it is by talking to people, engaging with people, and in a way that's a worry because our financial resources [are] going, which means our ability to support organizations to do it is reducing."[24]

The risk of making data collection, manipulation (through hackathons or other means), and presentation by citizens to decision-makers the de facto process for gaining civic voice is that these activities don't provoke action for change and, worse, may serve as excuses for further limiting collectively provided services. Rooney continued: "Along with probably every other city in the world there's the top-down, bottom-up tension. Bristol's got a very engaged sort of citizens. . . . But there's a real disconnect and I've heard . . . people say, 'Where's the power, where are decisions made? Why am I not told? Why don't I have a say?'"[25] Rooney's words echoed views expressed in my other interviews across Bristol: that citizens might install sensors, collect their data, discuss the meaning of the data, and open up a conversation, but these efforts might have little impact on strategic decision-making.

The Bristol Approach pilot unfolded in a context where investment in institutional resources (not always strategically made) had given way to organizational commitments to platform-based governance, including invitations to contribute data, knowledge, and expertise to civic processes. The city hasn't yet outsourced the damp inspector—the damp inspector has simply been replaced with a data-based process. Sensor information might solve the problem if only it were legitimate. The unevenness of the sensor data and the difficulty of bringing together sensor information, open data, and community-based knowledge complicate the seamless replacement of damp inspectors with sensors and bring out difficult questions about how and under what circumstances these forms of institutional legitimacy might be replaced with entrepreneurial or collaborative action. These frictions may well be more helpful in understanding how sensing data might work to address issues of justice than the notion that sensing might straightforwardly speak on behalf of the marginalized. As citizens grapple with the possibility of using data to address serious issues of inequality or injustice, these factors need to be carefully considered.

Friction may be inherent in commons. Resource-based commons, like pastures or fishing grounds, are governed by agreements between all of the beneficiaries and participants. Often researchers assume that the management of the commons depends on applications of consistent rules within relatively small, homogeneous communities to smooth or ease tension.[26] This isn't always the case. In some investigations of data commons, for example, researchers contrast the way that expansions of intellectual property create "anticommons" that produce conflictual relationships around the use of data, with the production and management of data commons providing an alternative.[27] Within scientific research, data sharing has expanded over the past few decades, sustained by large-scale institutional arrangements led by national research councils, such as the US National Academy of Sciences and the UK Research Councils. In medical research, patient data is often held in repositories that can be accessed by researchers, although this is changing. Google Artificial Intelligence spin-off DeepMind created a partnership with the Royal Free Hospital in London, where the company acquired a set of data on kidney patients in exchange for providing results of diagnostic tests. Despite efforts to privatize health data in the UK, there are still proposals to develop collaborative mechanisms to ensure collective benefits for medical data generated within the National Health Service.[28]

Writing about the ethics of contributions to data resources like biobanks, Barbara Prainsack and Alena Buyx argue that solidarity is expressed through action, not thought. In Bristol, there was little sense of a shared position or problem until sensing data started being collected, at which point, as Carolyn Hassan points out, "data defined the community . . . defined the way people thought about community," and "the community decided that those participating in the project were contributing to a shared resource."[29] Project participants not only contributed data but collaboratively budgeted the funds for design and installation of sensors and tried to ensure that the sensor hosts included community members who struggled to pay their energy bills. Stories of empowerment emerged: one participant who installed both a damp sensor and an electricity meter noticed that his bill did not correspond to the electricity as he had measured it. He petitioned to the electricity company and received compensation. The Bristol Approach created a repository of data on self-reported damp houses, which was correlated with open data on other factors like health, house prices, and everyday habits. Participants contributed photographs of their households and through workshops created and cemented relationships that created different ways to represent, visualize, and analyze the problems of damp. As a result, the municipal government has proposed changing licensing conditions for landlords who have damp properties.

Nevertheless, even this very small-scale example illustrates some of the difficulties in establishing frameworks of solidarity in relation to the use of data within commons, which may become more pronounced when and if data are used to seek justice in other contexts, like the ones I mentioned above. Some of the most inspiring stories from the Bristol Approach project were about individual, not collective, actions, like the story of the participant who succeeded in paying less for electricity. Other individual actions seriously challenged the idea of a shared data commons. One of the participants in the pilot opened a small business providing Frogbox damp sensors to a housing association that wanted to address the complaints they received about damp. The business model would have included a higher rate for landlords who wanted to keep the data private and a lower one if they shared the sensor readings with tenants.

Differential rates would have worked well for the entrepreneur but less well in sustaining a commons. At this point of tension, the project challenged the idea that a commons of sensor data could become a collective resource and speak for the people. Instead, this dynamic highlights inherent fractures and frictions. The commons was meant to be sustained by a set of protocols illustrating how to use the data. The protocols were at least partly established, however, by the missing figure of a damp inspector, who could have brought knowledge and experience to questions about whether a home had damp and who should be responsible for it. Instead, the debates about what damp data might mean and who could generate value from collecting it raised a set of broader questions about how sensor data is legitimated and what social practices develop around the commons.

THE POSSIBILITIES OF SOLIDARITY

It might be possible to use the Bristol Approach experience to reposition data commons as structures that support solidarity—as what philosopher Richard Rorty describes as "the imaginative ability to see strange people as fellow sufferers." The Bristol case also reveals that existing projects, relationships, and activist identities within cities create networks of support as well as create frictions. Solidarity, as Axel Honneth defines it, depends on reciprocity. Bioethicists Prainsack and Buyx define solidarity as "enacted commitments to accept costs to assist others with whom a person or persons recognize similarity in a relevant respect."[30]

The work of Prainsack and Buyx suggests ways that collecting and maintaining data resources can generate feelings of solidarity. First, solidarity, as they conceive of it, is axiological—focused on the potential to enact principles oriented toward particular outcomes—rather than normative: focused on ideal behaviors. Second, contributions to shared data resources can be seen as solidaristic when they are focused on aspects of similarity rather than difference, on what connects us rather than what separates us. Third, solidarity practices are embedded in notions of personhood that are relational and shaped by the relationships between people and their environments.[31] We can see the work of various groups in Bristol enacting many of these principles, since the various parts of the pilot process created new sites for action, new ways of defining shared issues with safe housing, and relationships between the different organizations interested in using technology in creative ways.

The Bristol case adds another quality of solidarity through data-based contributions: the experience of friction. The tensions produced in the struggle to create and maintain data commons made it possible to see the limitations of simply straightforwardly applying data to a problem and started to make people (even those within the local government) aware of the risks of seeing data collection as always good for its own sake. In contrast to the smooth optimization that constrains citizenship, friction slows down the process and requires negotiation, balance, and acceptance of similarity within difference.

It might be possible to focus on how collecting data destabilizes expertise without removing the necessity for it, producing an opening for considering how to build new frameworks and spaces where that legitimacy can be exercised. So, for example, in the platform framework for city governance that removes the damp inspector, what other legitimate actor might emerge? Could other participants in networks reposition themselves as legitimate in this way? Is there any way for the frictions and emergent solidarities to push past a vague idea that collecting data is a new form of civic participation?

GENERATIVE FRICTIONS

Bottom-up or community-driven data collection through citizen sensing promises new modes of effortless participation and new sources of voice. While the idea that what can be sensed only needs to be good enough to begin a conversation, looking at what happened in the citizen-sensing pilot in Bristol reveals that the relationships constructed between people and institutions with different interests are significant. Furthermore, this project suggests that the idea of sensing as presenting bottom-up knowledge may be unhelpful, since the knowledge may be contested even among citizens or contested between many different members of a network. (An example of the latter is the damp sensor knowledge that filled a gap where once an expert inspector had been.) There are tensions and connections here between a community using sensing to build knowledge of its environment and the difficulty of gaining institutional responses from a government forced to cut staff. These tensions can illuminate systemic problems, and this is perhaps one of the ways that the data commons can be transformative. The complexities of managing the commons and the systemic challenges that even small data-collection projects reveal may begin to make space for new relationships.

As the Bristol case study illustrates, in urban governance situations where fragmentation and privatization of urban assets (even informational assets) lead to greater appropriation and control of information, data commons can slow down, complicate, and open up these forms of power and control for debate. The promise—and practice—of building a commons can generate discussion, debate, and action. A kind of data solidarity might then emerge from the friction and contention around meaning, power, and social benefit, which could lay the basis for determining how to reduce datafication to a viable minimum.

CHAPTER FIVE

The Ends of Optimization

ARE or migratory birds might attract attention from birdwatchers, but crows, pigeons, and parakeets are more common in the London skies. These birds are expert foragers. They are highly adaptable. And parakeets, as indigenized migrants purportedly descended from escaped pets, are living illustrations of the shifts in London's climate over the past decades. Numerous and noisy packs of parakeets occupy parks, including leafy Richmond Park in West London, where they live alongside herds of deer. Parakeets are decried in local media as aggressive, noisy interlopers, dangerous migrants who are accused of crowding out locals or attacking their young. Local birds, too, like the many varieties of undistinguished little brown birds I can see in my backyard in London, are neighbors of mine, and I worry about the ones I don't see returning year after year. Along with so many other features of urban life, birdwatching and wildlife

spotting have been transformed both by the knowledge of environmental pressures, which has connected data about the numbers of animals and their endangered status, and by the cultures of participation that are part of citizen science and citizen-sensing activities. With the expansion of sensor and recording technologies citizens have been invited to contribute observations: beginning in 1998 the National Audubon Society in the United States has hosted annual citizen-science birdwatching events where people report observations of birds in their backyards or gardens, data that now combines with sensor-based data on the eBird system.

While these are ostensibly scientific endeavors that are expected to produce detailed data about bird presence and behavior, they are also sites where people come to terms with the realities of diminishing bird numbers and settings where data collection opens out the potential for collective grief over the transformation, reduction, and narrowing of the natural world. Even as I write these words and revisit birdwatching data-collection sites, I scroll over the old records for woodpeckers no longer seen in my part of London. There is no direct action to take in relation to bird observation—sometimes there is simply the pleasure of seeing another kind of being alive in the city or the grief at knowing you may never see a particular shape fly by again.

The world of animals produces another way of encountering the city, one where the kind of knowledge that might be produced or optimized through urban data flows is called into question. Urban foxes, for example, trigger the infrared sensors that are meant to work as burglar alarms. You can see this either as a malfunction of the burglar alarm or as an emergent property of its design, where the alarm also acts as a fox sensor. Such misrecognitions of living beings in data force themselves into our urban consciousness and disrupt the maintenance of a smoothly or perfectly managed world.

Queer theorist and artist Helen Pritchard calls these unexpected and emergent readings "feral data" and argues that the messiness of tracking animals defeats efforts at rendering them into a knowable whole. She describes how the feral cows of Hong Kong, whom the city government sees as a nuisance, have started to be monitored by radio collars. The location of cows, especially the "rebellious" locations that cows stray back to from the places where they have been relocated, becomes part of how they are defined as nuisances. At the same time, geolocating feral cows also makes it easier for people to organize themselves to provide food and care to the cows and to advocate for letting them live in the city rather than being forcibly relocated to exurban locations like rubbish dumps.[1] Ironically, although some care may be taken with these animals as a result of sensing their location, the same kinds of consideration and care are not always extended to people displaced or impacted by the expansion of smart-city projects, which are also being used as contemporary strategies of colonization and displacement.[2]

Especially when the paradoxes of sensing and knowing are kept in mind, the presence of animals provides a good way to start to question the seamlessness of urban information systems. Not only foxes are misrecognized. Geese were originally misrecognized as shopping carts in water quality sensing applications in Oxford. Thinking about misrecognized animals also helps us to start thinking about the way that people are misrecognized and about how the emergent properties of "design" strategies for data-optimized smart cities also cause misrecognition. The misrecognition holds some potential to use data to suggest or encourage other ways of being together in the city, but these will have political consequences. A look into some of the relationships between people, plants, and sensors in a community garden helps to identify these consequences and to assemble a theory of hybridizing knowledge that draws on sensing but is based in solidarity. Datafication can move away from being an aim in itself. Sensing knowledge can also provide another way for different kinds of knowledge or experience of the city to "be heard," and forms of knowledge commons that collect sense or knowledge may act as new ways "to speak"—perhaps by minimizing datafication while expanding other kinds of knowledge production.

ALTERNATIVE HISTORIES OF OPTIMIZATION

The dominant techno-systemic frame of the mid-twenty-first century is anchored in data-based optimization. This optimization draws on data collection and leverages this for platform-based control. In its logic it requires continuous expansion—of networks, platforms, and spheres of life to be datafied. Optimization, though, is not the same as smartness. Rather than opening out toward many different forms of knowledge, optimization narrows down to align data collection with the smoothing out of calculative processes. Making a system work faster or use a different subset of data is an internal optimization goal: improving a system on its own terms, not addressing what citizens might be concerned with in their everyday lives.

Smart cities, the corporate/commercial ones, give us only the narrowest possible ways to imagine or live in the urban world. There are other ways of thinking about sensing. Following from the acknowledgment that alternatives like the commons also generate friction comes the idea that frictions and tensions lead to transformations, to hybrid knowledge that combines the gaps and fissures in data and the experiences of people in cities, who are often living far from where they started out. Considering this combination creates space to reconsider where and how participation transforms cities: not through the expectation that technologies can optimally generate voice but through the frictions that emerge between technological knowledge and the relationships between people and other urban dwellers. Datafication can detract from the emerging intelligence of cities. Perhaps what I call minimum viable datafication is required to address this consequence, to create space for other kinds of knowledge.

Within the history of the big data optimized city, there are alternative ways of interpreting the potential for an augmented city—one where humans can use technology to better understand what lies beneath and beyond their own sensory capacity and to see the world differently. The ideas behind augmentation suggest that attention paid to flows, cycles, and systems might provide a different kind of engagement and embeddedness in place. This kind of knowledge is imagined as making the city "sentient"—capable of self-knowledge through reiteration and recalculation of data and information, but also lively in its own way. Now, the possibility of knowledge from sensors gestures at a fuller comprehension of ever more aspects of the city, from maps of trees to records of animals collected through citizen-science projects. Such projects suggest other potential avenues for participation— but they also reiterate the limitations of thinking of participation and civic action in techno-systemic terms.

Some features of the sentient city highlight the gaps between data and experience, which leave places for knowledge to develop in different ways. Writing about his vision of the sentient city, sociologist Nigel Thrift focuses on what geographer Doreen Massey calls the "throwntogetherness" of urban life: the way that various experiences and modes of feelings overlap, conflict, and interrelate, a concept that political theorist Jane Bennett also builds on when she sees political life as composed of encounters of "vibrant matter." Sensing doesn't smooth things out. As Thrift says, "Our perceptions of time and space are skewed in particular directions and although we have instruments that now allow us to see at least some other registers, we cannot produce an instant compass."[3] Still, he claims that the very things that make up the vision of the augmented city— the intensive connectedness, the flow of data, and the loosening of assumptions that data might need to be true or meaningful— can produce another way of knowing distinct from the optimized self-knowledge suggested by the visions of datafied cities.

MORE-THAN-HUMAN CITIES

Cities are part of the human-made world, but they are also more than human, not just because of the sensors and high-tech equipment that promise to make their complexity manageable. Cities are more than human also because of the animals and the plants that live with us. Increasingly, these fellow city residents are also becoming the subjects of sensing. The relationships between the different urban nonhumans show once again how computational and technological visions seep into the spaces that shape our knowledge. Trying to sense the nonhuman urban world provides some ways of directing our political attention that are unique to the augmented city, but these fail to account for the depth and complexity of knowledge. What is beyond the knowledge of technology, and how, paradoxically, can sensing systems show it to us?

Using environmental sensors to experience knowledge of others is one way for urban dwellers to encounter the natural world, as denuded and marginalized as it may have become. Thinking differently about sensing makes evident the potential for relationships of care and solidarity within situations of friction, as well as different conceptions of being-in-the world that result from processes of hybridization or encounters with others. Sentience slips away to make room for other things, and the techno-systemic focus on ontologies, circulation, and even breakdown can occlude possibilities for other kinds of engagements. The sensing city is not only a space that is represented or even representational. Equally, it is not only fragmented or broken. Even as things cease to work and begin to break down, new sites of potential are created for hybridizing knowledge. By itself, bottom-up knowledge production cannot transform how things are known about the city because it contributes to the same extractive and calculative dynamics of the big data optimized city. Other methods may be required to disassemble the colonizing influences of datafication, beginning with the use of sensing to perceive others.

BEYOND THE GOD-EYE: MONITORING THE ANIMAL OTHERS

Much of the dominant or canonical description of how urban spaces become augmented assumes that measurement or sensing creates some new knowledge to be managed.

On the contrary, sensing is not an extension of previous knowledge but another way of knowing. It is a contentious one, not only because the politics of sense data can reveal massive disparities between official and contextual experience (as we saw in the citizen monitoring of air quality) but also because glitches, failures, or uncertainties related to sensing reveal the complex texture of the relationships that characterize cities. As Bennett writes, "Theories of democracy that assume a world of active subjects and passive objects begin to appear as thin descriptions at a time when the interactions between human, viral, animal, and technological bodies are becoming more intense."[4]

For example, wireless sensor networks have now been used in remote monitoring of many kinds of living systems. In some cases, these recapitulate the assumptions that human perception provides the optimal way of seeing and understanding the world. An example is a project hosted by University College Cork and Libellium, a sensor technology company, for a wirelessly networked sensor for monitoring pollen growth inside beehives.[5] The hives rest on an electronic sensor that sends a message to the beekeeper's smartphone when the weight of the hives changes by more than a specific margin. Donna Haraway describes this sort of observation as the "god-eye" of objectivity, where we will make the bees knowable to us and we will allow something of their mystery to become comprehensible to us.[6] The amount of pollen can be shifted from something that bees sense or that a beekeeper feels to a set of data points on a mobile device, rendered through a set of technical protocols. This abstracts the sensations that link together the keeper and the bees. The promise is that the bees become knowable to the keeper—that this data, this knowledge, is what produces action, rather than the action of interacting with the hive to see how heavy it has become. We want to know more; we want to be able to transform sense into data to log the changes to the hive over time.

Something else is at work here: one possible response to the trauma of living not just as a human but as an urban human in the mid-twenty-first century, marked as it is by intensifying climate catastrophe and viral pandemics. In our current historical and cultural moment we suffer collectively from the ongoing trauma of the extinction of many of our fellow world-inhabitants and from disruptions to climate, land, and liberty. Urban dwellers in particular, given their obviously diminished contact with the living world, experience many of these shocks at a distance, since so much of their everyday life takes place in a setting without many plants or animals. The environmental shocks of climate change may appear distant, intransigent, and unknowable (except when they seep into quarantine restrictions aiming to prevent infections caused by the destruction of animal habitat), and as if any effective action were impossible. Climate change is a global problem but difficult to sense individually and locally, beyond an observation of changing weather. The extinction of perhaps a majority of species on earth is similarly difficult for urban dwellers without intimate everyday connections to animals to comprehend. So, in the same cities where utopian fantasies of connectivity promise a new frontier for civic expression and commercial profit, and where the intensification of capital takes its fullest expression in the datafication of everyday life, people wish to count bats and weigh bees, to tag foxes and map magpies. The ethical struggle between the abstraction of life through data and the experience of living together comes to the fore in the kinds of displacements demonstrated by sensor-based beekeeping projects, for example.

With the rise of the modern smart city, informational worlds have come to replace sensory or relational worlds— what design theorist and decolonial theorist Arturo Escobar calls the "pluriverse" of different being and different ways of seeing things.[7] Urban planning from above and urban life from below become concerned with the logistics of life, and technological frames and paradigms for citizenship channel creative civic energies toward technocratic engagements. All along, though, other ways of engaging have been not only possible but practiced. Behind the desire to transform beehive management into data lies an older, nonmodern wish to understand the bees, now intensified by the global (but not always situated or lived) knowledge of the threat of extinction for bees.

Faced with the global and individually ungraspable challenges of climate change and species extinction, people feel disempowered and depressed. One way of seeking to address this deep existential grief is by acknowledging its power. Environmental philosopher and indigenous elder Robin Wall Kimmerer offers the insight that humans might, in gratitude to the rest of creation, pay attention. She writes, "Paying attention to the more-than-human world does not lead only to amazement; it leads also to acknowledgment of pain. Open and attentive, we see and feel equally the beauty and the wounds, the old growth and the clear-cut, the mountain and the mine. Paying attention to suffering sharpens our ability to respond. To be responsible." Kimmerer's philosophy stresses the importance of human responsibility to the broader living world, the significance of grieving for the earth in order to regenerate it, and the importance of considering the flourishing of all, even when faced with the reality of displacement and destruction.[8] Paying attention is the least we might be expected to do as we strive for the grace to accompany fellow species toward their (and perhaps our) extinction; we might, perhaps, also strive for the will to imagine collective action to regenerate the earth.

FRAGILE SENSING: CONNECTED SEEDS

Since smart-city schematics represent the fullest extension and incursion of capitalism into everyday life, another hybridizing knowledge should take account of ways to push beyond the focus on the atomized individual and toward consideration of other perspectives—a position typical of decolonial theoretical perspectives like Escobar's. Such positions may become necessary in order to see schematics for what they are and to imagine how they might be otherwise. Key to this second transformative position is the idea of paying attention as a political act. This includes paying attention to the living world of animals and plants. In a community garden in East London, master gardeners who have come to the city from many different places explore the capacities and fragilities of sensing in relation to their own knowledge of growing.

London's Brick Lane acts as a mecca for tourists in search of curry, hosts young people drinking beer at picnic tables, and features a procession of walkers who gawk at the layered graffiti and increasingly expensive shops. Around the corner, backing onto a housing estate and nestled next to a train line and a playground is Spitalfields City Farm. Here there are animals, a wild garden, a field kitchen, and an incredible range of greenhouses, raised beds, covered seating areas, and indoor gardens. Sara Heitlinger, a design researcher and keen gardener, has been working here for eight years, experimenting with bringing different kinds of sensing technologies into the garden space. She began by conceiving of technologies that would help visitors to the garden understand what was happening in the beehives.[9] In 2015, Sara began working with other engineers and computer scientists on the Connected Seeds and Sensors project, which used connected Internet of Things sensors to gather environmental data at locations across the farm with the aim of investigating how technology could assist with food sustainability. With Heitlinger's support, I followed her project from its early stages to its conclusion, when data collected through environmental sensors, stories, recipes, and interviews was brought together into a library containing seeds collected from plants grown on the farm.

The Connected Seeds team used participatory design to determine what kinds of information should be added to its library. Seed libraries and commons are an ancient form of shared knowledge. Globally, seed libraries (like the one in Svalbard, Norway) hold examples of commercially important crops to secure their existence in an unknown future. The rise of corporate agribusiness and the increased commodification of seeds have put pressure on these kinds of physical and knowledge commons, and researchers like Heitlinger are trying to reinvigorate the commons by adding different types of knowledge to them, including sensor data. Heitlinger and her research team collected sensor data using custom-made temperature, humidity, and light sensors mounted in plastic cases installed indoors and outside near some of the crops. She also spent time interviewing gardeners, as well as asking them to photograph their gardens and write about the plants they cultivated.

Most of the gardeners at Spitalfields moved to London from other places in the world, bringing with them seeds, plants, and the knowledge of how to grow them. Migrant people might bring migrant plants like kodu, which is traditionally grown in South Asia, and callaloo, a leafy green often grown in the Caribbean. After being grown in a greenhouse for a few generations, the plants begin to hybridize. At Spitalfields volunteer gardeners hold knowledge about how to grow these plants in the conditions present in the garden. The Connected Seeds project followed fourteen Seed Guardians over a growing season. The researchers installed Arduino-based air temperature, humidity, soil temperature, and light monitors in the plots where the plants were grown. But these data were only a small part of what the project eventually tried to collect: the researchers also wanted to collect stories connected with the plants and reasons why the growers decided to grow them and collect and share seeds.

These stories, as well as the seeds, are displayed together in a portable seed library, where containers of seeds with QR codes on their lids can be lifted out and read. A spinner makes it possible to select different kinds of information, and a video monitor plays parts of video interviews. The interviews focus on why growers grow, how they feel in the garden, and how seeds connect to their history and heritage. The giant book-shaped library invites the visitor to play around a bit—it is fun to see different videos by turning the spinner and then looking at the seeds in the jars. Library members can borrow seeds and are expected to return seeds from the same plant species after their own growing season.

KNOWLEDGE COMMONS AND
HYBRIDIZING KNOWLEDGE

The Connected Seeds and Sensors project, like many of the other projects described in this book, put forward visions of a smart city that continue to engage with the dominant ways that sensing and data-based technology are conceived. In an essay for the project, Carl DiSalvo writes, "If we want Smart Cities to be something else, or to be something more . . . then we don't just need more design, what we need is a different design . . . sensing and automation, data mining and machine learning may have a role to play in diverse agricultures and community food systems. Perhaps even more provocatively, small-scale agriculture, gardening, foraging, and the like may be able to contribute insights into the building, use, and maintenance of Smart Cities."[10] While DiSalvo asks specifically how labor might be understood through projects like Connected Seeds, there are equally generative possibilities for rethinking sensing, data, and knowledge and challenging what DiSalvo calls the "zombie futures" of technology-driven visions of citizenship. It may be that the knowledge needed to fill gaps in the knowledge offered by sensor data provides an invitation to consider how knowledge in practice in places like community gardens may create local ways of knowing. These may have some resonance with philosophical perspectives generated outside the Western tradition, specifically in the tradition sometimes called "indigenous knowledge."

The library collects gardeners' stories that show how understanding from sensors connects with lived experience. Richard, a Seed Guardian and master gardener, reflects on having collected a season's worth of information and speculates in his interview with Heitlinger that "maybe if we'd had data from last year it would have been interesting to see [to what extent] this year things didn't grow so well."[11] But what do the data tell? The project's sensors were installed over only one growing season, and many of them collected data intermittently. One was stolen, along with a whole plot's worth of borlotti beans and a dozen beetroots. The sketchy sensor data, along with photographs of plants and story excerpts, were visualized in an interactive storyboard. Playing with the storyboard while revisiting my own photographs and memories from the garden, I find the sketchy and intermittent data only tendentially and tangentially connected to the kinds of knowledge circulating in the garden and woven into the seed library. Nevertheless, these data are valuable, perhaps because of their incompleteness. They highlight the ways that the knowledge of the gardeners exceeds the kind of data that can be collected through sensors during only a single season. As Lee Hester and Jim Cheney write in their discussion of North American indigenous epistemology in the Dakota tradition, which they gathered through years of work with Lakota elders, "Knowledge shaped by indigenous principles of epistemological method guarantees that knowledge is the result of deep and continuous communication between humans and the more-than-human world of which they are citizens."[12] If this kind of communication happens in the garden, though, it happens in a context where knowledge is always changing as a result of the movement of plants, people, and animals.

For the gardeners, who may lack formal citizenship in the UK but who possess relational knowledge of the garden space and the ways to make their migrant plants flourish, sensing highlights what can and cannot be known about the environment of the garden. Over seasons, the plants change as they adapt to the London climate, and the gardeners carry this shifting knowledge with them to the next growing season. Through recipes and stories, they create forms of knowledge that change over time. The knowledge is specific, place-based, but unbounded, since neither plants nor people nor conditions are fully stable. Yet the people are contributing to so much: to the reinvigoration of the soil, to a fruitful community, and to the stories, digital data, and shareable seeds of the seed commons.

Interviews with the fourteen Seed Guardians whose seeds are in the library present different reasons for saving seeds and growing food. One common thread is a sense of the importance of growing things as a way of dealing with displacement and finding belonging in a large global city. The fourteen guardians were born in six different countries. Basilia Gondo, seventy-two years old, describes herself as being from Zimbabwe and is one of a group of people who grow plants on the Zimbabwean Association plot in the garden. In her interview she says, "We grow our own food because it's got the taste we want, the taste which we grew up with. If we eat things from here the taste is not the same as we grew up with, and we get depressed." When Heitlinger asks about the relationship between growing and Basilia's heritage, she replies, "We are stressed because we are not home. We are out somewhere overseas. When we take pictures we send home they always laugh, 'Have you got a field in UK?' I said, 'Yeah, we have a field.' They said, 'Oh, we thought the UK was just full of cement and tired places.'" Another Guardian, Anwara Uddin, who lives in a small flat, grows runner beans and mustard in a plot on the communal lawn of her apartment block.

In her interview she says, "I come from Bangladesh. When I was small, I saw my mother growing these vegetables. I would help her. . . . When I came to this country, I missed everything. After twenty years I started gardening. I live in a flat, without a garden. I am happy now that I am growing my favorite plants. I feel better now."[13]

The Guardians bring experiences of growing from their different places of origin, but through their gardening practice they also come to know things about the other people whom they live with in London. Halema Begum, who gardens with Anwara and who also comes from Bangladesh, has lived in London for more than twenty-five years. Both she and Anwara talk in their interviews with Heitlinger about the way that gardening creates spaces for conversations with neighbors. "They just come and give me advice, or they just talk to me about themselves, their problems, and I'm just listening and doing my own work."[14]

The communal and social benefits of gardening, as well as the nascent political potential of gardens and other common urban spaces, are often celebrated. Researchers looking at Glasgow's community gardens saw the gardens as spaces of social transformation that held the potential for a new political practice. "Enabled by an interlocking process of community and spatial production, this form of citizen participation encourages us to reconsider our relationships with one another, our environment, and what constitutes effective political practice."[15] Of course, there is a risk of romanticizing gardens and gardening as pastoral correctives to urban alienation. What the stories and the connections in the seed library illustrate is that data, knowledge, and practice are all partial and in friction, creating space to hybridize. These frictions and this space for different kinds of knowledge to emerge around technological potential might be the most significant feature of the explorations at Spitalfields.

GAPS AND POTENTIAL BEYOND DATAFICATION

Some of the frictions concern the gaps in the sensor data collected over a single growing season. It might be easy to say of this project, as with the damp sensing, that the projects simply didn't create good-enough data and that better equipment would help to make the data a more significant part of the political voice of the gardeners. Another way of thinking about this might argue that "goodness" of data is beside the point. In Chapter 4 I discussed Jennifer Gabrys's sensing projects and the points of contention regarding the legitimacy of sensor data. In subsequent work she develops the concept of "data stories," arguing that the points of expression and contention that sensing data make possible do not necessarily only depend on the validity or calibration of data but are instead significant because of the narratives they permit.[16] So, for example, measurements of air quality that improve when a sensor is placed in a garden may make it possible to argue for the protection of gardens (which of course may generate many more benefits and outcomes than simply an improvement of air quality).

Something else is at work in the gaps in the construction of the knowledge commons in the Connected Seeds project. The gaps in the sensing data demonstrate the importance of the other kinds of knowledge about the adaptations required to effectively grow crops adapted for other climates, based on knowledge generated far away from where it is developed and practiced to grow things in new places and in new ways. They also demonstrate the deep meaning of the Guardians' participation in creating their neighborhood and their city. By bringing different plants and different ways of knowing, they physically and socially transform their patch of London.

Gardeners described the uncertainties of growing and the vagaries of the seasons and provided tips for growing unusual plants in the specific conditions of apartment living in inner London. Lutfun Hussain, who grows kodu, a South Asian gourd, describes how to start the seeds: "At home, end of March, I start planting seeds in a pot. I put it on top of the boiler where it is warm. When the seeds start to germinate, I put the pot on a sunny window. If you leave it on top of the boiler too long, the plant will grow too big and it won't survive." As Heitlinger points out in her writing about the project, "There were also challenges inside the gardens, typical of small-scale and community agriculture . . . crops failed due to pests, but also because of the vagaries of an uncertain climate. This unpredictability is set to increase with climate change. Therefore, it is ever more important for growers to have access to locally grown and adapted seed."[17] Seed is a resource for growth, but also a vector for hybridized knowledge.

Heitlinger and her partners devised this project to investigate ways of using technology to increase urban food sustainability. My observation and analysis of the project from its inception to conclusion highlight a different set of concerns—notably, the ways that this project, because of its community focus and inclusion of people, seeds, and knowledge, transcended automated data collection, suggesting a different way of conceiving of the significance of sensing.

Heitlinger poses these issues as questions of values, writing, "What values are we compromising if we automate small-scale and community gardening? [The gardeners] don't want to lose opportunities for face-to-face communication."[18] I see these tensions as invitations to focus on hybridizing knowledge and as indications that datafication in service of optimization can displace the potential for data and other types of knowledge to interconnect. Datafication can't occur without making some forms of information more valuable than others; collecting all of the data all of the time doesn't automatically open out new potentialities for knowledge creation; rather, it solidifies the interests of the actors who are most able to take advantage of it.

In contrast, the many types of knowledge held together by the context of the garden keep shifting. Ideas about relational ontologies and connections between ways of knowing and being include ethical and legal perspectives that emerge from both European philosophy and some of the indigenous knowledges of North America. These ideas may be just the sorts of hybrid ideas that illuminate the experiences of knowledge I'm describing here. They are, if you like, the antithesis of ideas about abstract, universally applicable measurement or sensing, the opposite of ideas of interchangeable infrastructure that could or should be used to manage any city. The garden at Spitalfields is extremely specific. It contains a particular combination of plants, light, heat, people, animals, particles, planter beds, and sensors not seen in any other garden in the world. The specific story of the gardeners at Spitalfields illustrates how territorial knowledge in our contemporary urban experience is a result of recombinations and hybridizations of knowledge and practice from many

originating territories and modes of thinking: these come together and transform in ways that exceed, but are brought
into relief by, the inadequate capacity of sensing to totally
capture this knowledge or to totally represent it visually.

ANOTHER ETHICS OF SENSE

Techno-systems thinking can seriously influence, perhaps even
distort, some of the qualities and actions that we might otherwise associate with citizenship. But technologies also transform
the ways that it's possible to think about knowing. Unfortunately, sensing, like the other smart-city technologies, is already
positioned in a certain way, in service of optimization and data-
based governmentality. I'd like to propose an ethics of sense that
draws on two key principles: solidarity forged in difference and
the political power of situated knowledge. The second idea has
come to prominence in discussions of decolonization of knowledge and positionings of indigenous wisdom within indigenous
and European philosophical traditions. It might seem difficult
to place it in the context of Global North cities built by the expansion of capitalism and its associated domination of other
ways of knowing and being, but my view is that it may be possible if undertaken with a sense of humility and an attention to
the instability of hybridity. Here it seems important to identify
my own sense of place: writing, myself, with the knowledge
gained from having spent the first decades of my life living as a
colonial settler on Plains Cree land and now residing in a global,
colonial city, I have my own particular consciousness of how
place and time inspire hybridity. As a loosely defined citizen of a
global city (though not of the nation it's in), I continue to carry
peculiarly grounded and hybrid perspectives—and wish to share

these here to see how and whether they grow. I theorize from my position, neither wishing to be nor capable of speaking on behalf of any of the indigenous thinkers cited here, and am mindful of the privilege that permits me to present these ideas.

Hybridizing global cities are territories of knowledge and constitutive relationships, with knowledge practices and attention proceeding from many cultural locations, not only the ones developing in European philosophy or social science. The frictions of sense, knowledge, and sensory data provide one small way of revealing these complexities, as well as one possible way past the narrowing-out of optimization. The aims of optimization are narrow, but they layer over a deeper and more complex undercurrent of counter-optimization, an undercurrent that attempts to link experience to territory and where, increasingly, sensing intertwines with other forms of knowledge.

Integrative perspectives on knowledge absorb the potential of sensed knowledge and also reject it as being the only possible truth. Dakota philosopher Vine Deloria Jr. identifies how some tribal traditions hold an expansive view of what could or should be known and how this knowledge is connected with land and territory: "In most [North American] tribal traditions, no data are discarded as unimportant or irrelevant. Indians consider their own individual experiences, the accumulated wisdom of the community that has been gathered by previous generations, their dreams, visions, and prophecies, and any information received from birds, animals, and plants as data that must be arranged, evaluated, and understood as a unified body of knowledge."[19]

To see and create forms of citizenship and participation that counter the seamless flows of data and its commercialized optimization requires a different kind of attention. I suggest that this attention builds on an ethic of nonreduction, of taking the world as an interconnected whole, where, as Lakota philosopher Suzanne Kite reiterates, the "ontological status of non-humans is not inferior to that of humans."[20] In indigenous philosophies that foreground relationality, nonhuman entities, including animals and others, are considered potential relations. Unlike other philosophies with assumptions about relationality, North American indigenous philosophies do not take abstraction or scaling-up as definitive values, a point that may be helpful in rethinking how sensing might be understood. "Relationality is rooted in context and the prime context is place," write four indigenous thinkers convened by Jason Lewis. To this Kimmerer adds the context of time, suggesting that real, focused attention on what is meaningful in one place solidifies meaningful arrangements of rights, responsibilities, and resources.[21] These arrangements are important not because locality or closeness are themselves valuable but because the kinds of situated knowledge produced through generations of experimentation have been subsequently dismissed, undermined, and forgotten.

These concepts of relationality, contextuality, and place are what ground indigenous forms of knowledge to the territory from which they proceed: knowledge of territory is developed in order to guide others for everyone's well-being. In deep knowledge of territory, rocks or trees or water or sky might be understood as being deeply integrated with all forms of life, and a good path forward would take into consideration these relationships and the energy required to sustain them into the future. The foundation of this future sustenance of relationships rests in an orientation of respectful attentiveness, as Kimmerer foregrounds. Such attentiveness can, and is, practiced anywhere, although the spiritual aspects associated with its practice have been the subject of genocidal suppression within colonial projects.

These philosophical traditions, still sometimes dismissed as anthropological curiosities from exotic or even extinct cultures, provide generative insight for seeing the sensing city differently. Since I am not an indigenous person, nor have I studied any of these philosophical traditions in detail, I introduce them here not because I wish to speak on behalf of the people who have created them but to generate a conversation that responds to other possibilities for sensing and to the trauma of a rapidly changing climate and global political and cultural instability. These currents of thoughts are "living and practiced by people with whom we all share reciprocal duties as citizens of shared territories."[22] Within the indigenous legal traditions in place in countries like Canada, New Zealand, and the United States, they represent legal orders in the territories to which they pertain. It would be disingenuous and dangerous to claim that ways of seeing the world that are embedded and proceed from thinking developed over time in particular places can be reappropriated and reapplied everywhere without attention to context. For those of us living in other landscapes and with other notions of time and space, these ideas provide different points of engagement with questions of knowledge and the more-than human that can open out away from existing perspectives.

Within this broad discussion, conversations and resonances occur between philosophers who are also thinking about sentient environments and other ways of knowing. Belgian philosopher Isabelle Stengers employs the concept of Gaia, which is similar to the Inuit concept of *Sila* (sometimes translated as "climate"). Métis anthropologist Zoe Todd describes *Sila* as "bound with life, with climate, with knowing and with the very existence of beings." An important organizing concept in the Arctic, its conceptual relative Gaia has been discussed since the 1960s by European philosophers as a way to understand ecological patterns as self-regulatory. In Stengers's recent work, which also concentrates on issues of shared cosmopolitical concern, she, too, advocates paying attention to the sentience and agency of nonhumans. Her Gaia is powerful and implacable (different somewhat from the power of the land that speaks in some indigenous philosophical and spiritual traditions), and there is no recourse to this power. Stengers writes, "We will have to go on answering for what we are undertaking in the face of an implacable being who is deaf to our justifications."[23]

Measurement and sensing undertaken for their own sake or with the aim of deriving and acting on completely objective knowledge will not create the means to address the shared crisis of extinction or climate change. Nor will preserving the environment as if it were separate from human life. Knowledge in and of place and interrelationships between different living and nonliving or human and nonhuman entities provide a counter to the capitalist tendency to use nature as material for development and exchange. This countervailing knowledge fragments, hybridizes, and reconstitutes in one global dynamic, even as colonial extraction of data and knowledge continue.[24] It also resists optimization.

HYBRIDIZING THE KNOWLEDGE COMMONS: MINIMUM DATAFICATION

The Connected Seeds library is intended to become a knowledge commons. Crucially, it contains more than seeds and stories of their hybridization. On a very small scale, the library presents a competing vision of how sensing and data can be employed in ways that confront and turn away from narrow data-based optimization. Like the data commons in Bristol, though, this commons has had to be negotiated among the Guardians and participants in the project. The farm has a grow-to-sell program, and this includes provision for the farm to sell rare or unusual seeds. Some participants, early in the project, worried that including too many seeds in the library might limit the money raised by growing rare seeds to sell. Once again, the promise of a knowledge commons is best made evident in the disagreements and difficulties in determining who and how it should be managed.

Building data commons, even when they are embedded in particular places and practices, does not automatically produce a kind of citizenship practice that might be recognized by conventional democratic theorists; it is not a matter of "add democracy and stir." Who gets to make trade-offs about what data go into a seed library, how many people are able to take seeds from it without returning them, how the stories and recipes can be interpreted in the absence of the people who contributed them—all these are difficult to balance. The Connected Seeds library was originally intended to be connected to the internet, but as the project unfolded, it became clear that connectivity would be burdensome and that the project needed to create a stand-alone library and a printed book to address the inevitable failure of technology. Crucially, these design decisions have resulted in a kind of minimum viable datafication that resists rendering all of the information about the seeds, stories, and relationships in the form of datafied outputs. Instead, resisting connectivity for the sake of it and presenting the stories and information in visual and audible stories, the library carries knowledge that can be taken forward in a number of different ways.

These small challenges, which can map outward to questions about big data and the management of larger or more complex forms of data commons, form part of an advocacy for minimum viable datafication. Part of the allure of volunteered or "implicitly participatory" data is that the challenges and complexities of different forms of knowledge that make up part of shared and lived spaces become narrowed into modes of production of value. Wendy Brown characterizes the relationship between neoliberal governance and citizenship by claiming that "governance is the consensus model of conduct integrating everyone and everything into a greater project with given ends. Governance replaces law with benchmarking . . . conflicting interests with stakeholders, political or normative challenges with a focus on the technical and practical. . . . Citizenship in its thinnest mode is mere membership," and "active citizenship is slimmed to tending oneself as responsibilized human capital."[25]

The gaps between data and knowledge as they are negotiated in the creation and maintenance of the seed commons push back at this narrowing of active citizenship by highlighting the gaps between what is known and what is measured, by helping to determine what should be subject to datafication—rather than what can be. They also highlight the hybridity of plants, people, and knowledge and suggest different ways of understanding the changes and adaptations that occur over time. Urban knowledge commons might be best understood as places and times where knowledge accumulates, where it changes and hybridizes as people and other living beings encounter each other as well as the technologies of smart cities.

This, too, is a form of participation, but one undertaken outside the framework of the ideal data citizen, who, having hoped to employ technology as a form of voice, finds these hopes compromised by a city whose smartness remains just outside, underneath, or in spite of optimization.

The Right to Minimum Viable Datafication

HE techno-systems thinking that influences smart-city projects extends the commodification of communication and personal data and defines good technological citizenship as supporting optimization. This ends up narrowing civic action to consumption of communication resources or to participation in streamlining data-based flows of information. However, networking, reinterpreting data, and sensing may also create spaces for a collective voice and novel forms of civic participation. The tensions in datafication show that power and agency are always at work in influencing who can speak, be heard, or act in relation to things that matter in the places they live.

The current and emerging features of the smart city may create small spaces for citizens to express themselves and to intervene in urban governance. Unfortunately, these are framed and structured in relation to political-economic arrangements supporting expanded data collection and consolidation of control over data analytics. These arrangements have been legitimated by both commercial and activist efforts at connecting cities to networks, optimizing through big data, and, more recently, promising to know the city through sensing. From both top down and bottom up, technology companies, cash-strapped governments, and enthusiastic tech-savvy activists have celebrated and legitimated optimizing and rendering more efficient ever more features of urban life. Optimization only works on a single aspect of a system, however, and the costs of optimization are often externalized, with consequences for people's lives in cities and their capacities to act as citizens. In particular, the equipment of cities with sensor systems and the big-data-optimization paradigm combine to make surveillance cheap for cities, externalizing the costs to people and communities. Yet civic voice still remains and can be produced in the gaps and at the points of disjuncture between these dominant positions—and in the spaces where it is possible to know otherwise or accept the unknowable value of urban life. Is this enough?

In this concluding chapter, I extend the critique of techno-systems thinking to two emerging smart-city proposals, the Alibaba City Brain project that has unfolded in Hangzhou, China, and the Sidewalk Labs project in Toronto, Canada. These projects share features that resonate with other smart-city proposals. They show the risks of pursuing strategies of optimization as goals above all, because seeking to create a smart city makes some knowledge and intelligence invisible. Technological citizens can show some ways to bring forward civic voices and can create different ways to claim rights and present critiques of the ways of knowing and modes of organizing that these smart proposals embed. Yet these may not be sufficient as modes of critique, because the smart cities seek to streamline and render efficient processes that eliminate some of the valuable friction that constitutes civic life. This by-product of optimization reappears both in challenges to smart-city projects and in imaginative reuse, breakdown, and reconstruction of its elements. When aligning citizen interests to the logic of optimization is not possible, power relations become skewed, and the smart city's optimizing capacities can turn toward the surveillant, oppressive, or totalitarian. I argue that it is no longer adequate only to advance normative critiques of smart-city projects and ask for more transparency or accountability. Instead, we must begin to recognize what the logics of optimization create, and strive for principles of minimum viable datafication—to collect data to develop functions in the present rather than to intensify future use—as a way to model an alternative approach.

ALIBABA CITY BRAIN

China's leading information company, Alibaba, began as an auction site similar to eBay. In the past decade its reach has expanded such that it now manages one of the most integrated information businesses in the world. The Alipay digital payment system, the Alibaba auction platform, and various other businesses now provide large amounts of data used to train Alibaba's machine-learning systems. In 2017, Alibaba partnered with the city government in Hangzhou, in eastern China, where its headquarters is located, to apply its artificial intelligence (AI) system to some of the well-identified issues with traffic management in the congested city. The City Brain project integrates data collected by Hangzhou's local government, including video from intersection cameras and GPS data on the locations of cars and buses, and subjects these to real-time analysis through its proprietary AI. The City Brain is managed by Alibaba, and although the governance arrangements for the data are unclear, it appears that its detailed analysis depends on constant access to Hangzhou city government data about the location and identification of vehicles. As the Chinese national government rolls out its Social Credit system—which links information about individuals, their credit scores, and financial status with information about their friends and relatives in order to enforce punitive controls over people whose network or data demonstrates a failure to abide by social norms—the level of individual vulnerability to smart-city control increases.

It is not difficult to imagine a City Brain system that would seek to preempt crime by arresting, detaining, or diverting people with undesirable social credit profiles. I share this example here to identify the many well-connected and well-established interests that seek to extend the optimizing capacities of smart cities in ways that actively seek to curtail individual or collective rights in the name of reducing disorder—which is, of course, another way of thinking about optimizing a system. In this case, the system may be the justice system, but increasingly, as other research has begun to show, various systems are folded into the same ways of thinking about and justifying data-based, computationally optimal cities. After a year of COVID-19 responses that have increased data-based control over people and their relationship to public space, these processes have accelerated. While Alibaba may be one of the world's most powerful data-based companies, similar drives toward optimization and similar models for the design and governance of smart cities continue to apply elsewhere—for example, in Canada and the United States.

SIDEWALK LABS

Sidewalk Labs, an urban development consultancy owned by Google's parent company, Alphabet, won a contract in 2017 to redevelop a section of the waterfront in Toronto, Canada. It was granted the contract by Waterfront Toronto, a special-purpose urban development organization supported by all three levels of government, but with an appointed, rather than elected, governing board. Following the model of public-private partnership, public entities like local governments often attempt to hold corporate actors to account (although they are often not able to achieve this). The arrangement, where an arm's-length entity with no explicit democratically mandated role makes decisions alongside a global company that is not held to follow local data protection laws, raises concerns for advocates and members of the public who are unsure who benefits from the Sidewalk project—and how. Although Sidewalk Labs backed away from the Waterfront Toronto project in June 2020, commentators have identified that the challenges raised by citizens to the proposal might be considered as a kind of testing ground for Alphabet, where citizen acceptance and interpretation of data governance are evaluated.[1]

One of the features of the Sidewalk Labs development was that it, like other greenfield smart cities, was meant to be built from the ground up. An expanse of waterside real estate in a former industrial area was set to be transformed into an area with housing for five thousand. Sidewalk Labs, as a platform provider, was meant to facilitate access to the latest technology for the residents (and access to the neighborhood's data for Sidewalk Lab partners and third-party application creators). In some early press coverage of the project, proponents described the smart city as working like a smartphone, in that the technology provider would be responsible for providing software for locating parking spaces or connecting smartphones to smart homes. This allocation of responsibility worried urban planners in Toronto: the experience of other greenfield smart cities, including Masdar City in the United Arab Emirates, Songdo in Korea, and PlanIT in Portugal, is that they incited investment, but the developer failed to build the projects in the manner promised.

Sensor technologies featured heavily in the plans for Sidewalk Toronto, and as I mentioned in Chapter 4, the plans originally specified that features would be unavailable to people who refused to register their personal data and accept monitoring across the entire Sidewalk development.[2] Many of the suggested innovations depended on collecting data on areas of life that had previously not been monitored, raising questions about who would benefit from this data collection, what consequences there might be for residents or citizens, and how the data would be governed on behalf of the population. Sidewalk Labs staged consultations, including some participatory design engagements. Some commentators thought Sidewalk Labs was potentially covertly gathering data and displacing effective consent.[3] One of the main considerations related to where the Sidewalk Labs data were to be stored and managed; given that Sidewalk Labs is an American company, there was no way to ensure that the data generated by the project would stay in Canada and remain subject to Canadian privacy laws. Press coverage revealed that Sidewalk Labs planned to sell location-based data from mobile phones that operated in Sidewalk Toronto even though a majority of

Canadians would oppose use of this data, even for local traffic planning. It is possible that this was one of the aspects that Alphabet wished to test out: exploring the extent to which citizens and various levels of government might oppose data collection or respond to different data governance proposals.[4]

Those who opposed Sidewalk Toronto critiqued the role of Sidewalk Labs in defining the shape and purpose of the neighborhood, with many of the innovations, such as self-driving electric cars, sensor-equipped sidewalks, and smart and responsive architecture, seemingly designed more to effectively collect and produce data than to respond to urban needs. Four members of the advisory board resigned in 2018, including Ann Cavoukian, the former privacy commissioner of the province of Ontario. Her involvement was meant to ensure that privacy would be at the heart of the smart-city design. Yet Sidewalk Labs reported that her expectations for privacy by design were unrealistic and suggested that these issues could be addressed by placing all of the data in a Civic Data Trust that would allow a third-party oversight board to control access to the data, which would not be directly owned by Sidewalk Labs. Data trusts, like the one that has been proposed to manage civic data in Barcelona, are usually established through consultation among the various entities that benefit.[5] As with data commons, the needs and expectations of different beneficiaries have to be balanced. Bianca Wylie, an activist vocally opposed to the governance arrangements of Sidewalk Toronto, argued that without involvement from civic groups in the management of the data trust, the trust could not be assumed to represent the public interest. The Toronto City Council approved the Sidewalk Labs project despite substantial civic opposition, and Alphabet only withdrew from its agreement after this approval had been granted.[6]

Sidewalk Labs claimed that its business model was to license the technology developed and tested in Toronto, rather than to profit from selling its data. However, the construction of the Civic Data Trust provided the potential to commercialize or to license access to the data once they had been collected. The concept of the Sidewalk Labs proposal was based on the development of a large commons of data, access to which could be sold, brokered, or provided in exchange for the development of new smart-city technologies. Sidewalk Labs and the Waterfront Toronto partners argued that the possibility of selling data from a data commons expanded the potential for innovation resulting from the potential to develop new and different kinds of technological applications based on the capacity to equip an entire new neighborhood with any and all of the sensors and data collection technologies that might be possible. However, this argument does not change the basic premise of the project: that urban problems are best addressed by equipping urban spaces with networked data and sensing technologies and by employing any information produced through this technological equipment as a resource for optimizing processes. As we have seen, optimization of this sort constrains civic action. It may also automate citizens' responses to stimuli, narrow their field of experience, and make technologically well-connected people subject to consistent nudges that make them better consumers but less connected to one another, less reflective, less active. Brett Frischmann and Evan Selinger view this as a form of "re-engineering humanity," and Shoshanna Zuboff has become concerned that the dynamics explored here, and in her more expansive work, are positioning surveillance capitalism in ways that become nearly impossible to resist.[7]

REDEFINING RIGHTS

I have shown in this book how rights to speak, listen, and be heard have been reinterpreted to justify and consolidate the dominant interpretations of smart cities and the modes of capitalist power that collapse citizenship into consumption of internet access or that legitmate platform-based neoliberal governance. Rights to communicate may be reappearing in new ways that can be understood from alternative ethical perspectives: a right to speak underpinned by real-time data may also be a way of speaking knowledge as it transforms.

The smart city in its data-based version measures everything and optimizes the processes that can be best represented through measurement. There are civic efforts to influence the consequences of such datafication by challenging the knowledge meant to be held by the data or the ontological power of organizing reality that smart cities promise, and these efforts can shift ways of thinking and acting. However, they still reinforce and reiterate the idea that a city is a system to be measured and made knowable and that data accrued as part of this measurement is a material that can be gathered and stocked.

As networked computational intermediaries promise to determine the best course of action in urban governance, and as issues come to be framed in certain ways "in order to be computable," two categories of response have emerged: normative responses that focus on justice in relation to calculability, and which employ strategies of transparency or accountability on automated processes, and critical responses that question the nature of data-based intermediation. Normative responses engage with the fact that calculation processes may be based on machine-learning processes that create outcomes that can't be anticipated in advance.[8] Many of these existing normative frameworks use existing frameworks such as notions of protected classes, abstraction, contextual integrity, and justice. However, thinking about data as relational may require other positions.

From a critical position, knowledge in the city encompasses far more than the data that optimize systems. It is constantly in motion, shifting as it moves along with people and comes to be produced in the relationships between people and others, including animals, plants, and—yes—technologies. It is not the processing through technology that renders data into knowledge or renders knowledge valuable or meaningful. It is the movement itself and the friction it produces that suggest that there is more to know than the dominant visions of the smart city will allow. Starting from this place, citizens can make room for the city to be a place of many changing ways to know and many methods of solving problems.

Optimization precludes and narrows the capacity for citizenship that uses technology to bring forward diverse knowledges in diverse ways and submits civic decision-making to narrow corporate interest. Viewed alongside the longer history of co-optation of changes in the communication environment by powerful commercial actors, the drive toward optimization of service delivery through data analytics is part of a trajectory toward private interests in the public sphere. Because this process appears to make space for the notion of civic voice and expression without the actual practice of either, it is difficult to unseat. It should be possible to address the fundamental shift in citizenship wrought by technology and discuss mechanisms of action rather than claims to the rights of citizenship. Here, we might wish to consider action against optimization.

Embedding techno-systems thinking in city governance subverts some of the practices that are essential for responding to the challenges of urban life in uncertain times: encountering and enduring tension and disagreement, or acknowledging the frictions that come from different opinions and divergent forms of knowledge. These sites of difference and tension, always in motion, are places where civic knowledge is produced and where the background work of democracy takes place. Across the many places and projects I visited, people worked hard with the resources provided through the dominant ideas of the smart city to extend and secure different kinds of rights, to challenge and redefine the kinds of knowledge constructed through official forms of data, and to show and respond to matters of concern and practices of community.

It is not possible to work outside the model of datafication or surveillance capitalism without an enormous political and social shift. The main concern should therefore be to examine how and in which way this shift might become possible. Legal theorist Julie Cohen suggests that one way to push back against datafication is to support the potential for "semantic discontinuity" where it is impossible or very difficult to link together different types of data and draw immediate conclusions from them.[9]

As we have seen in this book, the call for discontinuity should be extended to include not only semantic discontinuity but epistemological and ontological discontinuity. Liberty resides not only in the freedom from surveillance that Cohen and Zuboff argue for but also in the capacity to know about things in a different way, to disagree about interpretations, to have some aspects of life, individual or collective, remain somewhat unknowable.[10] These rights to discontinuity secure certain forms of liberty—of thought, of interpretation, of collective action taken from a perspective that could be oppositional. These rights to discontinuity also hold open the capacity to maintain competing or nonintersecting priorities as equally important: a necessary friction.

To secure the democratic potential of community organizations and motivated individuals, people who care about the kinds of belonging made possible in hybridizing cities as well as in the many other contexts where we live among others must resist the technocratic impulse that dilutes rights and narrows them down to opportunities to consume. This impulse, spurred on by discourses, protocols, and forms of techno-systems thinking, may be prevalent in the urban places where these ideas have taken hold, but the counterarguments emerge in many places, not all of which have traditionally been understood in relationship to rights. This lack of input needs redress: perspectives from relational ethics, including some forms of indigenous and cosmopolitical philosophy, identify how responsibility and care are also part of what makes up the "right to have rights."[11]

In the name of innovation, smart cities have focused on technological frameworks that produce optimization. Accompanied first by enthusiasm about the capacity of networking and then by promises of inclusion, accountability, and greater participation by virtue of access to data, these frameworks have produced and reproduced techno-systems thinking. At its worst, this mode of thinking can legitimize surveillance and the constant efforts to render into data ever more aspects of daily urban life, with the assumption that this trajectory will create efficiency in delivering urban systems or effective social oversight, as in the case of the Alibaba City Brain. Efforts like the Sidewalk Labs data trust and other attempts to assess the ethical risks and benefits of sensor-based data extraction may succeed in making smart-city processes more transparent. However, this transparency may not solve the questions about how data is made meaningful and who benefits from this meaning-making.

As we have seen, sensor data can tie into processes where power is consolidated. Civic interpretations need to make space within the institutional arrangements that are most common, where data are either a stock of material awaiting analysis or personal property. Enforcing and enhancing rights to privacy are important—these rights may need to be reconsidered as applying not only to individuals but to groups of people whose shared data speaks to a matter of concern, as the contentions around the Bristol damp-data commons illustrate. As suggested by the process of hybridizing knowledge, citizenship in relation to sensing and data technologies can also emerge in the gaps and alongside the glitches where automated data collection fails; citizens can develop and create knowledge that belongs to particular places. These practices also help to situate citizenship, helping to demonstrate ways that civic action transcends universality and to identify ways to reconceive citizenship as a set of practices that seek justice, equity, and also spaces for flourishing relationships between people and others. How should these practices and spaces be created?

How can outcomes associated with authoritarian applications of integrating sensing be thwarted? Dealing with regulatory frameworks like that of the Alibaba City Brain or Sidewalk Labs in Toronto usually means insisting on transparency, accountability, or space for citizens to reinterpret data or make different epistemological claims using institutionally collected data. Unfortunately, such approaches don't modify the embedding of technological frames into urban life. One way to redress smart-city optimization could be to approach urban design and experience from a perspective of minimum viable datafication. We can acknowledge that data collection, including sensing data, continues to be enmeshed in circuits of power and influence and seek to minimize, rather than constantly extend, the kinds of data that are collected. Heitlinger and her coauthors explore a similar concept, identifying a "right to the sustainable smart city."[12] Such calls for rights are not surprising; indeed, they are pressing. Not only does datifying the economy have tremendous social consequences; it also has environmental ones, for the energy required to store and process digital data impacts climate and development in the cities and towns where data centers are located. Designing for minimum viable datafication in urban areas would need to begin from a consideration of what is already known and how this knowledge can be meaningfully connected with urban decisions. Certainly, sensor-based data collection, the creation and maintenance of data commons, and the acknowledgment that digital data-based knowledge cannot be holistic need to be part of this effort, but equally it should be possible to recast urban decision-making as not necessarily dependent on datafication.

Minimum viable datafication, then, is a necessary but not sufficient condition for facilitating relationships between people, governments, and spaces of urban encounter and securing and expanding communication rights. It is not necessary to assume that digital data is a store of value or a material for training future automated systems. This does not mean never collecting data; it does not mean never employing AI to parse patterns, or sensing devices to contribute to decision-making. It means following a strategy of minimizing, rather than maximizing, this kind of data, and it means seeking to employ decision-making strategies that may appear to be more costly on the surface but that leave space for different kinds of knowledge, as well as for data to decay over time, for frictions to be identified and addressed, and for different forms of democratic participation and accountability, including but not limited to data audit, sensing citizenship, and autonomous networking. It means leaving room for determining what can be known, claimed, and acted upon outside, against, and within the data.

FINAL WORDS, NOT QUITE OPTIMISTICALLY

Against a global backdrop of climate unpredictability, increased risks of global viral pandemics, the securing of borders amid mass migration, and political movements that draw on fear of others and promote protection of ostensibly scarce resources, cities could become many things: sites of sanctuary and spaces of environmental and communitarian resilience, efficient machines for living that host more people in less space and require management by smart machines, or dangerous spaces of division threatened by food shortages and inconsistent infrastructure. Cities will probably always be all of these, with the ways that they are understood in part driven by the expectations attached to them. The vision and promise of the smart city aligns the goals of optimizing connectivity, data collection, and sensing processes with civic actions that sustain them. Yet alongside and underneath this operationalization of citizenship lie other ways to sense, know, and relate. Minimum viable datafication is one pragmatic strategy for holding open the space for rights to discontinuity, for competing and non-intersecting priorities, that are all part of the dynamic life of a city. The right to discontinuity lets people care both about seamless traffic flow and traffic safety, both about expert judgment and civic participation, both about accountability and expertise. We should be able to retain rights and capacities to hybridize—to transform and evolve our ways of knowing as the world we inhabit continuously becomes less certain and less easy to narrow and optimize.

Notes

INTRODUCTION. TECHNOLOGY, CITIZENSHIP, AND
FRAMEWORKS OF THE SMART CITY

1. Taylor, 2004.
2. Mansell, 2012.
3. Brown, 2015.
4. Harvey, 2012; Couldry, 2012.
5. Stark, 1999; Benkler, 2006.
6. Castells, 2002; Friedman, 2005.
7. Nagy and Neff, 2015.
8. Padovani and Calabrese, 2014.
9. Mayer-Schönberger and Cukier, 2013; Van Dijck, 2014, 198.
10. Dean, 2005.
11. Dean, 2005, 199; Dencik, 2011.
12. Open Data Institute, 2017.
13. Couldry and Powell, 2014.
14. Gabrys, 2017.
15. Tsing, 2012, 147; OxFlood, presentations at UK IGF [Internet Governance Forum], 2015.
16. Tsing, 2012.
17. Holston and Appadurai, 1999.
18. Ostrom et al., 1994.
19. Frischmann, Madison, and Strandburg, 2014.

CHAPTER I. NETWORK ACCESS AND THE
SMART CITY OF CONNECTIVITY

1. Cohill and Kavanaugh, 1997.
2. Marx, 2000; Mosco, 2005.
3. Sassen, 1994 ; Castells, 1996; Appadurai, 1990.
4. Holston and Appadurai, 1999, 200; Isin and Ruppert, 2015.
5. Soja, 1999; Sorkin, 1992.
6. Raboy and Shtern, 2010.
7. Gurstein, 2003; Dean, 2005, 55; Lunt and Livingstone, 2011; Livingstone, Lunt, and Miller, 2007.
8. Shade, 2010; Government of Canada, 1996.
9. Greenfield and Kim, 2013; Kitchin, 2014; Mansell, 2014.
10. Greenfield and Kim, 31.
11. Bar and Galperin, 2005; Tapia and Ortiz, 2008.
12. At time of writing in 2020 the network still operates in a few locations on a best-effort basis, although the local government ceased to invest in it after 2006–7, when it began to invest in the municipally owned gigabit internet, which launched in 2014.
13. Based the author's online survey conducted in February 2007, as well as network mapping and ethnographic interviews conducted in February 2007.
14. Barron, 2013.
15. Coleman, 2012.
16. Medosch, 2014.
17. Original French: "C'est principalement un club de geek, ah, je pense que c'est un club de passionnés."
18. Original French: "On est une belle gng . . . il y a du beau monde ici."
19. Posting to the Île San Fil volunteer list, February 2005.
20. Powell, 2008; Warner, 2002.
21. Daniel Lemay interview, December 6, 2007. Original French: "C'est comme on a créé une chaine de production, on a répéter le modèle industriel. . . . La problème c'est qu'il n'y a pas vraiment des buts nobles. . . . En dedans il ya une problème de gouvernance. Les gens avec les projets artistiques étaient toujours les 'outsiders.'"
22. Île Sans Fil user, interview, November 5, 2005.
23. Herman, 2010, 197; Medosch, 2018.
24. Byrum, 2019.
25. Byrum, 2019.

CHAPTER 2. DATA CITIES AND VISIONS OF OPTIMIZATION

1. Van Dijck, Poell, and de Waal, 2018, 8.
2. Mayer-Schönberger and Cukier, 2013.
3. Kitchin, 2014; Leszczynski, 2016.
4. Van Dijck, 2014, 198.
5. Livingstone, Lunt, and Miller, 2007; Pilkington, 2019.
6. Amoore, 2013, 9; Ruppert, 2013, 38.
7. Poon, 2009.
8. Medina, 2011.
9. Kitchin, Lauriault, and McArdle, 2015, 13–14.
10. Kitchin, Lauriault and McArdle, 2015, 17; Halpern, 2017, paragraph 19.
11. Gillespie, 2013, 168.
12. Amoore and Piothukh, 2015.
13. Plantin, 2018.
14. Gillespie, 2010.
15. Van Dijck and Poell, 2013.
16. O'Reilly, 2011.
17. Maltby, n.d.; Balka, 2015; Gillespie, 2013, 8.
18. Pasquale, 2015.
19. Fourcade and Healy, 2017; Sadowski and Pasquale, 2015.
20. Cisco, n.d.
21. Siemens, n.d.
22. Urban Engines [Archived], n.d.; Urban Engines, 2016.
23. DataSift, 2017; Recorded Future, n.d.
24. Mackenzie, n.d.; Criteo, 2017.
25. Meaney, 2017.
26. Mackenzie, 2015, 432; McQuillan, 2015.
27. Mackenzie, 2015, 440.
28. Mackenzie, 2015, 443; Amoore, 2013, 67, 61.
29. Langley and Leyshon, 2016.
30. Osborne and Rose, 1999.
31. For more information on the projects see FixMyStreet (2017), https://www.fixmystreet.com; CycleStreets (2017), https://www.cyclestreets.net.
32. Gabrys, 2016.
33. Andrejevic and Burdon, 2014, 6, 24.

CHAPTER 3. ENTREPRENEURIAL DATA CITIZENSHIPS, OPEN
DATA MOVEMENTS, AND AUDIT CULTURE

1. Dreyfus, 1987.
2. Robertson and Travaglia, 2015.
3. Kitchin, Lauriault, and McArdle, 2015, 12–13.
4. Amoore, 2016; Amoore, 2013.
5. Bates, 2012; Bates, 2014.
6. Peña Gangadharan and Niklas, 2019.
7. Baack, 2015; Powell, 2015.
8. The tree map and the representations of cycle traffic are available at http://ingrid.geog.ucl.ac.uk/~ollie/misc/bookdata/southwark_trees_v2.pdf (courtesy Oliver O'Brien); see also a map of Bikeshare in London by OOMap, https://bikesharemap.com/london/timeline/#/12.555139390504332/-0.1197/51.5021/.
9. The definition is available at an Open Government Data site maintained by Joshua Tauber, https://opengovdata.org/.
10. Dencik, 2019.
11. Noveck, 2015; Schrock and Shaffer, 2017.
12. Conservative Party (UK), 2010.
13. Noveck, 2015.
14. Scott, 1998, 78.
15. GOV.UK, n.d.
16. Open Data Institute, n.d.
17. Open Data Institute, n.d.
18. O'Reilly, 2011.
19. Bates, 2012.
20. Baack, 2015, 4; Noveck, 2015.
21. Baack, 2015, 4.
22. Research assistance on this chapter was provided by Karen Morton.
23. Grenfell Action Group, n.d.; Feigenbaum, 2018.
24. Bruno, Didier, and Vitale, 2014, 213; Genel, 2013; Bruno, Didier, and Vitale, 2014, 208.
25. Davies, interview, 2012.
26. Davies, interview, 2016.
27. Davies, interview, 2012.
28. Bloom, interview, 2016.
29. Bloom, interview, 2016.
30. Bloom, 2013, 261.

31. Levine, 2007; Bloom, 2013, 264.
32. Schrock, 2018; Irani, 2019.
33. Davies, interview, 2016.

CHAPTER 4. RETHINKING CIVIC VOICE IN POST-NEOLIBERAL CITIES

1. Couldry, 2012; Couldry and Powell, 2014.
2. Breen et al., 2015.
3. Isin and Ruppert, 2015; Balestrini et al., 2017.
4. Graham and Marvin, 2002.
5. Zannat and Choudhury, 2019.
6. Sieber and Haklay, 2015, 127; Air Quality Egg, n.d.; Citizen Sense, 2013.
7. Gabrys, 2014, 38.
8. Cardoso and O'Kane, 2019; Lachman, 2020.
9. Foster and McChesney, 2014.
10. Tsing, 2011, xii.
11. The Bristol Approach, n.d.
12. Balestrini et al., 2017; Ideas for Change, n.d.; KWMC, n.d.
13. Rooney, 2016.
14. Hassan, 2016.
15. Ostrom, 1994.
16. Balestrini, 2016.
17. Balestrini, 2016.
18. Cuff, Hansen, and Kang, 2008, 28.
19. Tenney and Sieber, 2016, 104.
20. Mattern, 2017; Hollands, 2008; Gillespie, 2013.
21. GovLab, n.d.
22. Balestrini, 2016.
23. Balestrini, 2016.
24. Rooney, 2016.
25. Rooney, 2016.
26. Hess and Ostrom, 2005.
27. Fisher and Fortmann, 2010.
28. Information Commissioner's Office, 2017; Balestrini et al., 2017.
29. Prainsack and Buyx, 2017; Hassan, 2016; Balestrini et al., 2017.
30. Rorty, 1989, xvi; Honneth, 1991; Prainsack and Buyx, 2017, 5:2.
31. Prainsack and Buyx, 2017.

CHAPTER 5. THE ENDS OF OPTIMIZATION

1. Pritchard, 2013.
2. Datta, 2018.
3. Massey, 2005; Bennett, 2009; Thrift, 2014, 2.
4. Bennett, 2009, 108.
5. Libelium, n.d.
6. Haraway, 1988.
7. Escobar, 2018.
8. Kimmerer, 2014, 20; Kimmerer, 2013.
9. Heitlinger, 2017.
10. DiSalvo, cited in Heitlinger, 2017, 24.
11. Heitlinger, 2018.
12. Hester and Cheney, 2001, 325.
13. Heitlinger, 2018, 34, 14, 37.
14. Heitlinger, 2018, 41.
15. Crossan et al., 2016, 952.
16. Gabrys, 2016.
17. Heitlinger, 2018, 12.
18. Heitlinger, 2018, 12.
19. Deloria, cited in Hester and Cheney, 2001.
20. Lewis et al., 2018, 12.
21. Lewis et al., 2018; Kimmerer, 2013.
22. Todd, 2016.
23. Stengers, 2011; Todd, 2016, 5; Stengers, 2011, 47.
24. Todd, 2016; Couldry and Mejias, 2019.
25. Brown, 2011, 218, 211.

CONCLUSION. THE RIGHT TO MINIMAL
VIABLE DATAFICATION

1. Lachman, 2020.
2. Cardoso and O'Kane, 2019.
3. Mattern, 2020.
4. Kofman, 2019; Georgiou, 2018.
5. Morozov and Bria, 2018.
6. Oliver, 2019.
7. Frischmann and Selinger, 2018; Zuboff, 2019.
8. Barocas and Selbst, 2016.
9. Cohen, 2012.

10. Cohen, 2012; Zuboff, 2019.
11. Isin and Ruppert, 2015.
12. Heitlinger, Bryan-Kinns, and Comber, 2019.

Bibliography

Air Quality Egg. n.d. "Air Quality Egg—Science Is Collaboration." Accessed November 1, 2019. https://airqualityegg.com/home.

Amoore, Louise. 2013. *The Politics of Possibility: Risk and Security Beyond Probability*. Durham, NC: Duke University Press.

Amoore, Louise. 2016. "Biometric Borders: Governing Mobilities in the War on Terror." *Political Geography* 25 (3): 336–51.

Amoore, Louise, and Volha Piothukh. 2015. "Life Beyond Big Data: Governing with Little Analytics." *Economy and Society*, 1–26. https://doi.org/10.1080/03085147.2015.1043793.

Andrejevic, Mark, and Mark Burdon. 2014. "Defining the Sensor Society." *Television and New Media* 16 (1): 19–36. https://doi.org/10.1177/1527476414541552.

Appadurai, Arjun. 1990. "Disjuncture and Difference in the Global Cultural Economy." *Theory, Culture, and Society* 7 (2–3): 295–310.

Baack, Stefan. 2015. "Datafication and Empowerment: How the Open Data Movement Re-articulates Notions of Democracy, Participation, and Journalism." *Big Data and Society* 2 (2). https://doi.org/2053951715594634.

Balestrini, Mara. 2016. Interview. October.

Balestrini, Mara, Yvonne Rogers, Carolyn Hassan, Javi Creus, Martha King, and Paul Marshall. 2017. "A City in Common: A Framework to Orchestrate Large-Scale Citizen Engagement Around Urban Issues." In *CHI '17: Proceedings of the 2017 CHI Conference on Human*

Factors in Computing Systems, May 2017: 2282–94. https://doi. org/10.1145/3025453.3025915.

Balka, Ellen. 2015. "Mapping the Body Across Diverse Information Systems: Shadow Bodies and How They Make Us Human." In *Boundary Objects and Beyond: Working with Leigh Star*, edited by Geoffrey Bowker, Stefan Timmermans, Adele Clarke, and Ellen Balka. Cambridge, MA: MIT Press.

Bar, François, and Hernan Galperin. 2005. "Bar, François, and Hernan Galperin. 'Geeks, Cowboys, and Bureaucrats: Deploying Broadband, the Wireless Way.'" In *The Network Society*, edited by Manuel Castells and G. Cardoso, 269–87. Oxford: Oxford University Press.

Barocas, Solon, and Andrew D. Selbst. 2016. "Big Data's Disparate Impact." *California Law Review* 104: 671.

Barron, Anne. 2013. "Free Software Production as Critical Social Practice." *Economy and Society* 42 (4): 597–625.

Bates, Jo. 2012. "'This Is What Modern Deregulation Looks like': Co-optation and Contestation in the Shaping of the UK's Open Government Data Initiative." *The Journal of Community Informatics* 8 (2): 1–20.

Bates, Jo. 2014. "The Strategic Importance of Information Policy for the Contemporary Neoliberal State: The Case of Open Government Data in the United Kingdom." *Government Information Quarterly* 31 (3): 388–95.

Benkler, Yochai. 2006. *The Wealth of Networks: How Social Production Transforms Markets and Freedom*. New Haven: Yale University Press.

Bennett, Jane. 2009. *Vibrant Matter: A Political Ecology of Things*. Durham, NC: Duke University Press.

Bloom, Greg. 2013. "Towards a Community Data Commons." *Beyond Transparency: Open Data and the Future of Civic Innovation*, 255–70. Washington, DC: Code for America Press.

Bloom, Greg. 2016. Interview.

Breen, Jessica, Shannon Dosemagen, Jeffrey Warren, and Mathew Lippincott. 2015. "Mapping Grassroots: Geodata and the Structure of Community-Led Open Environmental Science." *ACME: An International E-Journal for Critical Geographies* 14 (3), 849–73.

The Bristol Approach. n.d. "The Bristol Approach to Citizen Sensing." n.d. Accessed November 1, 2019. https://www.bristolapproach.org/.

Brown, Wendy. 2015. *Undoing the Demos.* New York: Zone Books.

Bruno, Isabelle, Emmanuel Didier, and Tommaso Vitale. 2014. "Statactivism: Forms of Action Between Disclosure and Affirmation." *Partecipazione e Conflitto: The Open Journal of Sociopolitical Studies* 7 (2): 198–220.

Byrum, Greta. 2019. "Building the People's Internet." *Urban Omnibus.* October 2, 2019. https://urbanomnibus.net/2019/10/building-the-peoples-internet/.

Cardoso, Tom, and Josh O'Kane. 2019. "Sidewalk Labs Document Reveals Company's Early Vision for Data Collection, Tax Powers, Criminal Justice." *The Globe and Mail,* October 30, 2019, online edition, sec. News.

Castells, Manuel. 1996. "The Space of Flows." *The Rise of the Network Society* 1: 376–482.

Castells, Manuel. 2002. *The Internet Galaxy: Reflections on the Internet, Business, and Society.* Oxford: Oxford University Press.

Cisco. n.d. "Cisco Smart + Connected City Parking: Helping Cities, Citizens, Local Businesses, and Enforcement Agencies." Accessed November 5, 2019. https://www.cisco.com/c/dam/en_us/solutions/industries/us_government/resources/govconnection-smartcities.pdf: Cisco.

Citizen Sense. 2013. "Air Quality Egg and the Makers." *Citizen Sense* (blog). August 27, 2013. https://citizensense.net/air-quality-egg/.

Cohen, Julie E. 2012. *Configuring the Networked Self: Law, Code, and the Play of Everyday Practice.* New Haven: Yale University Press.

Cohill, Andrew Michael, and Andrea Kavanaugh. 1997. *Community Networks—Lessons from Blacksburg, Virginia.* Norwood, MA: Artech House.

Coleman, E. Gabriella. 2012. *Coding Freedom: The Ethics and Aesthetics of Hacking.* Princeton, NJ: Princeton University Press.

Conservative Party (UK). 2010. "Invitation to Join the Government of Britain: The Conservative Manifesto 2010." https://general-election-2010.co.uk/2010-general-election-manifestos/Conservative-Party-Manifesto-2010.pdf.

Couldry, Nick. 2012. *Media, Society, World: Social Theory and Digital Media Practice.* Malden, MA: Polity.

Couldry, Nick, and Ulises A. Mejias. 2019. *The Costs of Connection: How Data Is Colonizing Human Life and Appropriating It for Capitalism.* Stanford, CA: Stanford University Press.

Couldry, Nick, and Alison Powell. 2014. "Big Data from the Bottom Up." *Big Data and Society* 1 (2). https://doi.org/2053951714539277.

Criteo: The Advertising Platform for the Open Internet. 2017. "Commerce Marketing Where Everyone Wins." https://www.criteo.com.

Crossan, John, Andrew Cumbers, Robert McMaster, and Deirdre Shaw. 2016. "Contesting Neoliberal Urbanism in Glasgow's Community Gardens: The Practice of DIY Citizenship." *Antipode* 48 (4): 937–55.

Cuff, Dana, Mark Hansen, and Jerry Kang. 2008. "Urban Sensing: Out of the Woods." *Communications of the ACM* 51 (3): 24–33. https://doi.org/10.1145/1325555.1325562.

DataSift. 2017. "DataSift. Human Data Intelligence." http://datasift.com.

Datta, Ayona. 2018. "The Digital Turn in Postcolonial Urbanism: Smart Citizenship in the Making of India's 100 Smart Cities." *Transactions of the Institute of British Geographers* 43 (3): 405–19.

Davies, Tim. 2012. Interview.

Davies, Tim. 2016. Interview.

Dean, Jodi. 2005. "Communicative Capitalism: Circulation and the Foreclosure of Politics." *Cultural Politics* 1 (1): 51–74.

Dencik, Lina. 2011. *Media and Global Civil Society.* London: Springer.

Dencik, Lina. 2019. "Situating Practices in Datafication—from Above and Below." In *Citizen Media and Practice: Currents, Connections, Challenges*, edited by Hilde C. Stephansen and Emiliano Treré, 243–56. London: Routledge.

Dreyfus, Hubert. 1987. "From Socrates to Expert Systems: The Limits of Calculative Rationality." *Bulletin of the American Academy of Arts and Sciences* 40 (4): 15–31.

Escobar, Arturo. 2018. *Designs for the Pluriverse: Radical Interdependence, Autonomy, and the Making of Worlds.* Durham, NC: Duke University Press.

Feigenbaum, Anna. 2018. "Data-Based Story-Telling." American Association of Geographers Conference, New Orleans, Louisiana, USA, April 29, 2018.

Fisher, Joshua B., and Louise Fortmann. 2010. "Governing the Data Commons: Policy, Practice, and the Advancement of Science." *Information and Management* 47 (4): 237–45.

Foster, John Bellamy, and Robert W. McChesney. 2014. "Surveillance Capitalism: Monopoly-Finance Capital, the Military-Industrial Complex, and the Digital Age." *Monthly Review* 66 (3): 1.

Fourcade, Marion, and Kieran Healy. 2017. "Seeing Like a Market." *Socio-Economic Review* 15 (1): 9–29. https://doi.org/10.1093/ser/mww033.

Friedman, Thomas L. 2005. *The World Is Flat: A Brief History of the Twenty-First Century.* New York: Farrar Straus Giroux.

Frischmann, Brett M., Michael J. Madison, and Katherine Jo Strandburg. 2014. *Governing Knowledge Commons.* Oxford: Oxford University Press.

Frischmann, Brett, and Evan Selinger. 2018. *Re-engineering Humanity.* Cambridge: Cambridge University Press.

Gabrys, Jennifer. 2014. "Programming Environments: Environmentality and Citizen Sensing in the Smart City." *Environment and Planning D: Society and Space* 32 (1): 30–48.

Gabrys, Jennifer. 2016. *Program Earth.* Minneapolis: University of Minnesota Press.

Genel, Katia. 2013. "L'autorité des faits: Horkheimer face à la fermeture des possibles." *Tracés: Revue de Sciences humaines,* no. 24 (May): 107–19. https://doi.org/10.4000/traces.5658.

Georgiou, Myria. 2018. "Imagining the Open City: 189 (Post-) Cosmopolitan Urban Imaginaries." In *The Routledge Companion to Urban Imaginaries,* edited by Christoph Lindner and Miriam Meissner, 187–201. London: Routledge.

Gillespie, Tarleton. 2010. "The Politics of 'Platforms.'" *New Media and Society* 12 (3): 347–64.

Gillespie, Tarleton. 2013. "The Relevance of Algorithms." In *Media Technologies,* edited by Pablo Boczkowski and Kirsten Foot. Cambridge, MA: MIT Press. Retrieved from http://governingalgorithms.org/wp-content/uploads/2013/05/1-paper-gillespie.pdf.

Government of Canada. 1996. Building the Information Society: Moving Canada into the 21st Century (May 23, 1996).

GovLab. n.d. "The Governance Lab." Accessed November 1, 2019. https://www.thegovlab.org/.

GOV.UK. n.d. "Eric Pickles 'Shows Us the Money.'" Accessed November 4, 2019. https://www.gov.uk/government/news/eric-pickles-shows-us-the-money-as-departmental-books-are-opened-to-an-army-of-armchair-auditors.

Graham, Steve, and Simon Marvin. 2002. *Splintering Urbanism: Networked Infrastructures, Technological Mobilities and the Urban Condition.* London: Routledge.

Greenfield, Adam, and Nurri Kim. 2013. *Against the Smart City.* Part 1: *The City Is Here for You to Use.* New York: Do Projects.

Grenfell Action Group. n.d. "Grenfell Action Group." Accessed November 4, 2019. https://grenfellactiongroup.wordpress.com/.

Gurstein, Michael. 2003. "Effective Use: A Community Informatics Strategy Beyond the Digital Divide." *First Monday* 8 (12).

Halpern, Orit. 2017. "Hopeful Resilience." *e-Flux Journal.* April 19. https://www.e-flux.com/architecture/accumulation/96421/hopeful-resilience/.

Haraway, Donna. 1988. "Situated Knowledges: The Science Question in Feminism and the Privilege of Partial Perspective." *Feminist Studies* 14 (3): 575–99.

Harvey, David. 2012. *Rebel Cities: From the Right to the City to the Urban Revolution.* London: Verso.

Hassan, Carolyn. 2016. Interview. November.

Heitlinger, Sara. 2017. "Talking Plants and a Bug Hotel: Participatory Design of Ludic Encounters with an Urban Farming Community." PhD disssertation, Queen Mary University of London.

Heitlinger, Sara. 2018. *Connected Seeds.* EPSRC. London: Queen Mary University of London.

Heitlinger, Sara, Nick Bryan-Kinns, and Rob Comber. 2019. "The Right to the Sustainable Smart City." In *Proceedings of the 2019 CHI Conference on Human Factors in Computing Systems,* 287: 1–287:13. New York: ACM. https://doi.org/10.1145/3290605.3300517.

Herman, Andrew. 2010. "'The Network We All Dream Of': Manifest Dreams of Connectivity and Communication; or, Social Imaginaries of the Wireless Commons." *The Wireless Spectrum: The Politics, Practices, and Poetics of Mobile Media.* Toronto: University of Toronto Press. 187–98.

Hess, Charlotte, and Elinor Ostrom. 2005. "A Framework for Analyzing the Knowledge Commons: A Chapter from Understanding

Knowledge as a Commons: From Theory to Practice." *Libraries' and Librarians' Publications*, Paper 21.

Hester, Lee, and Jim Cheney. 2001. "Truth and Native American Epistemology." *Social Epistemology* 15 (4): 319–34. https://doi.org/10.1080/02691720110093333.

Hollands, Robert G. 2008. "Will the Real Smart City Please Stand Up? Intelligent, Progressive or Entrepreneurial?" *City* 12 (3): 303–20.

Holston, James, and Arjun Appadurai. 1999. "Introduction: Cities and Citizenship." In *Cities and Citizenship*. Vol. 1. Durham, NC: Duke University Press.

Honneth, Axel. 1991. *The Critique of Power: Reflective Stages in a Critical Social Theory*. Translated by K. Baynes. Cambridge, MA: MIT Press.

Ideas for Change. n.d. "Making Sense." Accessed November 1, 2019. https://www.ideasforchange.com/makingsense.

Information Commissioner's Office (UK). 2017. "Royal Free—Google DeepMind Trial Failed to Comply with Data Protection Law." July 3, 2017. https://ico.org.uk/about-the-ico/news-and-events/news-and-blogs/2017/07/royal-free-google-deepmind-trial-failed-to-comply-with-data-protection-law/.

Irani, Lilly. 2019. *Chasing Innovation: Making Entrepreneurial Citizens in Modern India*. Vol. 22. Princeton, NJ: Princeton University Press.

Isin, Engin, and Evelyn Ruppert. 2015. *Being Digital Citizens*. London: Rowman and Littlefield International.

Kimmerer, Robin Wall. 2013. *Braiding Sweetgrass: Indigenous Wisdom, Scientific Knowledge, and the Teachings of Plants*. Minneapolis: Milkweed Editions.

Kimmerer, Robin Wall. 2014. "Returning the Gift." *Minding Nature* 7 (2): 18–24.

Kitchin, Rob. 2014. "The Real-Time City? Big Data and Smart Urbanism." *GeoJournal* 79 (1): 1–14.

Kitchin, Rob, Tracey P. Lauriault, and Gavin McArdle. 2015. "Knowing and Governing Cities Through Urban Indicators, City Benchmarking and Real-Time Dashboards." *Regional Studies, Regional Science* 2 (1): 6–28.

Kofman, Ava. 2019. "Google's Sidewalk Labs Plans to Package and Sell Location Data on Millions of Cellphones." *The Intercept* (blog).

January 28. https://theintercept.com/2019/01/28/google-alphabet-sidewalk-labs-replica-cellphone-data/.

KWMC: Knowle West Media Centre. n.d. "REPLICATE." Accessed November 1, 2019. https://kwmc.org.uk/projects/replicate/.

Lachman, Richard. 2020. "Sidewalk Labs' City-of-the-Future was a Stress Test We Needed." *Policy Options.* May 28. https://policyoptions.irpp.org/magazines/may-2020/sidewalk-labs-city-of-the-future-in-toronto-was-a-stress-test-we-needed/.

Langley, Paul, and Andrew Leyshon. 2016. "'Platform Capitalism: The Intermediation and Capitalization of Digital Economic Circulation.'" *Finance and Society* 2 (August). https://doi.org/10.2218/finsoc.v3i1.1936.

Lemay, Daniel. 2007. Interview. December 6.

Leszczynski, Agnieszka. 2016. "Speculative Futures: Cities, Data, and Governance Beyond Smart Urbanism." *Environment and Planning A: Economy and Space* 48 (9): 1691–1708. https://doi.org/10.1177/0308518X16651445.

Levine, Peter. 2007. "Collective Action, Civic Engagement, and the Knowledge Commons." In *Understanding Knowledge as a Commons,* edited by Elinor Ostrom and Charlotte Hess, 247. Cambridge, MA: MIT Press.

Lewis, Jason Edward, Noelani Arista, Archer Pechawis, and Suzanne Kite. 2018. "Making Kin with the Machines." *Journal of Design and Science* (July). https://doi.org/10.21428/bfafd97b.

Libelium. n.d. "Case Studies." Accessed November 2, 2019. http://www.libelium.com/resources/case-studies/.

Livingstone, Sonia, Peter Lunt, and Laura Miller. 2007. "Citizens, Consumers and the Citizen-Consumer: Articulating the Citizen Interest in Media and Communications Regulation." *Discourse and Communication* 1 (1): 63–89.

Lunt, Peter, and Sonia Livingstone. 2011. *Media Regulation: Governance and the Interests of Citizens and Consumers.* London: Sage.

Mackenzie, Adrian. 2015. "The Production of Prediction: What Does Machine Learning Want?" *European Journal of Cultural Studies* 18 (4–5): 429–45. https://doi.org/10.1177/1367549415577384.

Mackenzie, Adrian. n.d. "Programming Subjects in the Regime of Anticipation: Software Studies and Subjectivity." *Subjectivity* 6 (4): 391. https://doi.org/10.1057/sub.2013.12.

Maltby, Paul. n.d. "A New Operating Model for Government." Policy Lab (blog). GOV.UK. Accessed November 5, 2019. https://openpolicy.blog.gov.uk/2015/03/17/a-new-operating-model-for-government/.

Mansell, Robin. 2012. *Imagining the Internet: Communication, Innovation, and Governance.* Oxford: Oxford University Press.

Mansell, Robin. 2014. "Empowerment and/or Disempowerment: The Politics of Digital Media." *Popular Communication* 12 (4): 223–36.

Marx, Leo. 2000. *The Machine in the Garden: Technology and the Pastoral Ideal in America.* Oxford: Oxford University Press.

Massey, Doreen. 2005. *For Space.* London: Sage.

Mattern, Shannon. 2017. "A City Is Not a Computer." *Places Journal,* no. 2017(July).https://placesjournal.org/article/a-city-is-not-a-computer/.

Mattern, Shannon. 2020. "Post-It Note City." *Places Journal,* 2020 (February). https://placesjournal.org/article/post-it-note-city/.

Mayer-Schönberger, Viktor, and Kenneth Cukier. 2013. *Big Data: A Revolution That Will Transform How We Live, Work, and Think.* Boston: Houghton Mifflin Harcourt.

McQuillan, Dan. 2015. "Algorithmic States of Exception." *European Journal of Cultural Studies* 18 (4–5): 564–76. https://doi.org/10.1177/1367549415577389.

Meaney, Andrew. 2017. "Hard Shoulder: Using Behavioural Nudges to Reduce Congestion." *Oxera: Compelling Economics* (blog). https://www.oxera.com/agenda/traffic-behavioural-nudges-reduce-congestion/.

Medina, Eden. 2011. *Cybernetic Revolutionaries: Technology and Politics in Allende's Chile.* Cambridge, MA: MIT Press.

Medosch, Armin. 2014. "The Rise of the Network Commons, Chapter 1 (Draft)." *The Rise of the Network Commons.* Armin Medosh's blog. *The Next Layer.* https://webarchiv.servus.at/thenextlayer.org/node/1231.html.

Medosch, Armin. 2018. "The Network Commons and the City as Project and Utopia." *International Journal of Electronic Governance* 10 (2): 120–37. https://doi.org/10.1504/IJEG.2018.093834.

Mignolo, W. 2011. "Epistemic Disobedience and the Decolonial Option: A Manifesto." *Transmodernity: Journal of Peripheral Cultural Production of the Luso-Hispanic World* 1 (2).

Morozov, Evgeny, and Francesca Bria. 2018. "Rethinking the Smart City." *Democratizing Urban Technology*. New York: Rosa Luxemburg Foundation.

Mosco, Vincent. 2005. *The Digital Sublime: Myth, Power, and Cyberspace*. Cambridge, MA: MIT Press.

Nagy, Peter, and Gina Neff. 2015. "Imagined Affordance: Reconstructing a Keyword for Communication Theory." *Social Media + Society* 1 (2). https://doi.org/2056305115603385.

Noveck, Beth Simone. 2015. *Smart Citizens, Smarter State: The Technologies of Expertise and the Future of Governing*. Cambridge, MA: Harvard University Press.

Oliver, Joshua. 2019. "Sidewalk Labs Reaches Smart-City Deal with Toronto." *Financial Times*. October 31. https://www.ft.com/content/9cd15bcc-fbff-11e9-a354-36acbbbod9b6.

Open Data Institute. 2017. "Smart Cities Overview." *The ODI*. https.//theodi.org.

Open Data Institute. n.d. "The ODI—Open Data Institute."Accessed November 4, 2019. https://theodi.org/.

O'Reilly, Tim. 2011. "Government as a Platform" Innovations: Technology, Governance, Globalization 6 (1) 13–40. https://doi.org/10.1162/INOV_a_00056.

Osborne, Thomas, and Nikolas Rose. 1999. "Governing Cities: Notes on the Spatialisation of Virtue." *Environment and Planning D: Society and Space*. 17 (December): 737–60. https://doi.org/10.1068/d170737.

Ostrom, Elinor, Roy Gardner, James Walker, James M. Walker, and Jimmy Walker. 1994. *Rules, Games, and Common-Pool Resources*. Ann Arbor: University of Michigan Press.

Padovani, Claudia, and Andrew Calabrese. 2014. *Communication Rights and Social Justice: Historical Accounts of Transnational Mobilizations*. London: Springer.

Pasquale, Frank. 2015. *The Black Box Society: The Secret Algorithms That Control Money and Information*. Cambridge, MA: Harvard University Press.

Peña Gangadharan, Seeta, and Jędrzej Niklas. 2019. "Decentering Technology in Discourse on Discrimination." *Information, Communication, and Society* 22 (7): 882–99.

Pilkington, Ed. 2019. "Digital Dystopia: How Algorithms Punish the Poor." *The Guardian*, October 14, 2019, sec. Technology. https://

www.theguardian.com/technology/2019/oct/14/automating-poverty-algorithms-punish-poor.

Plantin, Jean-Christophe, Carl Lagoze, Paul N. Edwards, and Christian Sandvig. 2018. "Infrastructure Studies Meet Platform Studies in the Age of Google and Facebook." *New Media and Society* 20 (1): 293–310. https://doi.org/10.1177/1461444816661553.

Poon, Martha. 2009. "From New Deal Institutions to Capital Markets: Commercial Consumer Risk Scores and the Making of Subprime Mortgage Finance." *Accounting, Organizations and Society* 34 (5): 654–74. https://doi.org/10.1016/j.aos.2009.02.003.

Powell, Alison B. 2008. "WiFi Publics: Producing Community and Technology." *Information, Communication, and Society* 11 (8): 1068–88.

Powell, Alison B. 2015. "Open Culture and Innovation: Integrating Knowledge Across Boundaries." *Media, Culture, and Society* 37 (3): 376–93.

Prainsack, Barbara, and Alena Buyx. 2017. *Solidarity in Biomedicine and Beyond.* Vol. 33. Cambridge: Cambridge University Press.

Pritchard, Helen. 2013. "Thinking with the Animal Hacker, Articulation in Ecologies of Earth Observation." In *A Peer Reviewed Journal about Back When Pluto Was a Planet, The Reinvention of Research as Participatory Practice (Transmediale)*, edited by C. Anderson and G. Cox. Berlin: Transmediale.darc.

Raboy, Marc, and Jeremy Shtern. 2010. *Media Divides: Communication Rights and the Right to Communicate in Canada.* Vancouver: UBC Press.

Recorded Future. n.d. "Threat Intelligence Powered by Machine Learning." Accessed November 5, 2019. https://www.recordedfuture.com/.

Robertson, Hamish, and Joanne Travaglia. 2015. "Big Data Problems We Face Today Can Be Traced to the Social Ordering Practices of the 19th Century." *LSE Impact Blog.* Politics of Data Series. October 13. https://blogs.lse.ac.uk/impactofsocialsciences/2015/10/13/ideological-inheritances-in-the-data-revolution/.

Rooney, Kathryn. 2016. Interview. November.

Rorty, Richard McKay. 1989. *Contingency, Irony, and Solidarity.* Cambridge: Cambridge University Press.

Ruppert, Evelyn. 2013. "Rethinking Empirical Social Sciences." *Dialogues in Human Geography* 3 (3): 268–73. https://doi.org/10.1177/2043820613514321.

Sadowski, Jathan, and Frank Pasquale. 2015. "The Spectrum of Control: A Social Theory of the Smart City." *First Monday* 20 (7). https://ssrn.com/abstract=2653860.

Sassen, Saskia. 1994. *Global City*. Princeton, NJ: Princeton University Press.

Schrock, Andrew. 2018. *Civic Tech*. Long Beach: Rogue Academic Press.

Schrock, Andrew, and Gwen Shaffer. 2017. "Data Ideologies of an Interested Public: A Study of Grassroots Open Government Data Intermediaries." *Big Data and Society* 4 (1). https://doi.org/2053951717690750.

Scott, James C. 1998. *Seeing Like a State: How Certain Schemes to Improve the Human Condition Have Failed*. New Haven: Yale University Press.

Shade, Leslie. 2010. "Access." In *Media Divides: Communication Rights and the Right to Communicate in Canada*, edited by Marc Raboy and Jeremy Shtern, 120–44. Vancouver: UBC Press.

Soja, Edward. 1999. "Thirdspace: Expanding the Scope of the Geographical Imagination." In *Human Geography Today*, edited by Doreen Massey, John Allen, and Philip Sarre, 260-78. Cambridge, UK: Polity.

Sieber, Renée E., and Mordechai Haklay. 2015. "The Epistemology(s) of Volunteered Geographic Information: A Critique." *Geo: Geography and Environment* 2 (2): 122–36.

Siemens. n.d. "Smart Cities." n.d. Accessed November 5, 2019. https://new.siemens.com/global/en/company/topic-areas/intelligent-infrastructure.html.

Sorkin, Michael, ed. 1992. *Variations on a Theme Park: The New American City and the End of Public Space*. New York: Hill and Wang.

Stark, David. 1999. "Heterarchy: Distributing Intelligence and Organizing Diversity." In *The Biology of Business: Decoding the Natural Laws of Enterprise*, edited by John Henry Clippinger III, 153–79. San Franciso: Jossey-Bass.

Stengers, Isabelle. 2011. *Cosmopolitics Ii*. Minneapolis: University of Minnesota Press.

Tapia, Andrea, and Julio Angel Ortiz. 2008. "Deploying for Deliverance: The Digital Divide in Municipal Wireless Networks." *Sociological Focus* 41 (3): 256–75.

Taylor, Charles. 2004. *Modern Social Imaginaries*. Durham, NC: Duke University Press.

Tenney, Matthew, and Renée Sieber. 2016. "Data-Driven Participation: Algorithms, Cities, Citizens, and Corporate Control." *Urban Planning* 1 (2): 101–13.

Thrift, Nigel. 2014. "The 'Sentient' City and What It May Portend." *Big Data and Society* 1 (1). https://doi.org/10.1177/2053951714532241.

Todd, Zoe. 2016. "An Indigenous Feminist's Take on the Ontological Turn: 'Ontology' Is Just Another Word for Colonialism." *Journal of Historical Sociology* 29 (1): 4–22.

Tsing, Anna Lowenhaupt. 2011. *Friction: An Ethnography of Global Connection.* Princeton, NJ: Princeton University Press.

Tsing, Anna Lowenhaupt. 2012. "Unruly Edges: Mushrooms as Companion Species: For Donna Haraway." *Environmental Humanities* 1 (1): 141–54. https://doi.org/10.1215/22011919-3610012.

Urban Engines. 2016.

Urban Engines [Archived]. n.d. "Your Key to the City." For Cities/Urban Engines. Accessed November 5, 2019. https://web.archive.org/web/20140605063403/http://www.urbanengines.com:80/cities.

van Dijck, José. 2014. "Datafication, Dataism and Dataveillance: Big Data Between Scientific Paradigm and Ideology." *Surveillance and Society* 12 (2): 197–208.

van Dijck, José, and Thomas Poell. 2013. "Understanding Social Media Logic." *Media and Communication* 1 (1). https://doi.org/10.12924/mac2013.01010002.

van Dijck, José, Thomas Poell, and Martijn de Waal. 2018. *The Platform Society.* Oxford: Oxford University Press. https://www.oxfordscholarship.com/view/10.1093/oso/9780190889760.001.0001/oso-9780190889760.

Warner, Michael. 2002. "Publics and Counterpublics." *Public Culture* 14 (1): 49–90.

Zannat, Khatun E., and Charisma F. Choudhury. 2019. "Emerging Big Data Sources for Public Transport Planning: A Systematic Review on Current State of Art and Future Research Directions." *Journal of the Indian Institute of Science* 99: 601–19. https://doi.org/10.1007/s41745-019-00125-9.

Zuboff, Shoshana. 2019. *The Age of Surveillance Capitalism: The Fight for a Human Future at the New Frontier of Power.* New York: Profile Books.

Index

access. *See* network access
activists and activism: against
  commodification of data, 79;
  communication commons and, 32,
  49–50, 52–53, 101–4; free
  information infrastructures of,
  39–42; government roles for,
  105–6; platformed logics and, 80;
  right to communicate, 13–14;
  standards of data commissioning
  and, 98–101; techno-systems
  thinking of, 100–101, 164. *See also*
  open data movements
air quality sensors, 113–14
algorithms, 61, 62, 64. *See also*
  cybernetic systems; machine
  learning
Alibaba, 164–66, 174–75
Allied Media Project (Detroit), 51
Alphabet (platform company), 67,
  167. *See also* Sidewalk Labs
  project
Amoore, Louise, 58, 73, 82
Andrejevic, Mark, 78
animals: extinction of species,
  143–44, 159; human control of
  nature and, 19–20; indigenous
  peoples, relationship with, 156;
  knowledge from human relationship with, 172; misrecognition of
  feral data, 20, 136–37; monitoring

of, 22–24, 141–45; relationships
  between, 140–41; urban dwellers
  lacking contact with, 143
attention as political act, 144–45,
  156–58
Audubon Society, 136

Balestrini, Mara, 109, 121, 124
Balka, Ellen, 64
Bates, Jo, 91–92
Bath: Hacked (group), 93–97
beehive monitoring, 142–44, 146
Beer, Stanford, 60
Begum, Halema, 151
Bennett, Jane, 140, 142
Berlin, 39–42, 49, 52
biases, 82–83, 93, 111, 113
big data optimized cities, 7–11;
  anticipatory meaning-making in,
  69–71, 82; citizenship, implications
  for, 71–75; corporate actors'
  platform strategies in, 65–69;
  cybernetic control in, 59–62;
  data-based value and, 62–65,
  139–40; datafication in, 57–59, 61;
  entrepreneurship in, 77–79;
  platformization of, 16–17, 55–56,
  104–5; public sector and civic
  organizations in, 75–77, 104; smart
  cities compared to, 58; splintering
  of services in, 16, 111–16, 126, 134;